817/98

Ink into Bits

A Web of Converging Media

Charles T. Meadow

The Scarecrow Press, Inc.
Lanham, Md., & London
1998

SCARECROW PRESS, INC.

Published in the United States of America
by Scarecrow Press, Inc.
4720 Boston Way
Lanham, Maryland 20706

4 Pleydell Gardens
Kent CT20 2DN, England

British Library Cataloguing in Publication Information Available

Library of Congress Cataloging-in-Publication Data

Meadow, Charles T.
Ink into bits : a web of converging media / Charles T. Meadow. p. cm.
Includes bibliographical references and index.
ISBN 0-8108-3507-X (cloth : alk. paper). — ISBN 0-8108-3508-8 (pbk. : alk. paper).
1. Mass media—Technological innovations. 2. Information technology.
3. Books and reading. I. Title.
P96.T42M4   1998   98-16854
302.23—DC21   CIP

⊖™ The paper used in this publication meets the minimum requirements of American National Standard for Information Sciences—Permanence of Paper for Printed Library Materials, ANSI Z39.48–1984.
Manufactured in the United States of America.

To those who came before and helped to show the way,
my brother, Stephen Meadow, and my oldest remaining
relatives, Aunts Esther Saxe and Shirley Walker,
this book is dedicated with love.

Contents

Figures

Preface

This book presents my own views of the media and communication issues that are emerging out of the shift from print toward electronics in publishing. These issues affect what we now call books, journals, and magazines and their readers, as well as motion pictures and television shows and their viewers. The users of these media include nearly everyone in our culture. My intent in writing the book is to discuss the issues in non-technical terms, and without the unseemly hype that often attends "futurology." The changes that may be coming, are coming, or may already be here will have a profound impact on the worlds of publishing, schools, libraries, and everyday reading. Reading, even for recreation, in turn, has a profound impact on how our culture exists and is passed on to our children.

It is not necessary to be any sort of expert to understand this book. I assume that only people who are interested in books, journals, magazines, newspapers or the visual media, for themselves, will read it. I generally use the broadest meaning of *book* because the media are converging and the distinctions among what we now think of as books, magazines, television programs, or motion pictures will very likely disappear.

In 1980, I and my family were on a crowded train bound from Shef-field to London, England. People were standing in the aisles. Near us was a family who had immigrated from India, the father a physician practicing in nearby Rotherham. They had a daughter, about three or four years old, whose seat was a suitcase in the aisle and who entertained our end of the coach by singing English nursery or Mother Goose songs, in a rich Yorkshire accent. *She* was going to grow up English. I grew up on the same songs, having no more English blood than that little girl. But, these are the kinds of things that bind us together. And we learn them from books, read or sung to us before we ourselves can read.

While it is not my intent to cry havoc, I hope to bring to my readers a sense that big things are in the offing and that it behooves any literate person to be aware of the issues and perhaps to exert some influence on their outcome. The issues do not have the immediacy of wars, pestilence, and poverty. But in the long run, they might be as important in terms of what they do to our civilization. We can all have some effect on what they do and we can exert influence simply by our buying habits.

I will discuss the kinds of changes we are seeing or are soon likely to see in the recording and transmission of information to homes, schools, and offices, with emphasis on the first two, on how these developments will affect ordinary people. There is no technological knowledge required. I offer brief explanations of some technology, without detailed descriptions. But there is thinking required. I do not do that for you. My aim is to explain likely effects, or to get readers thinking about effects. I will un-abashedly present my own opinions, but I will generally try to get read-ers forming their own, rather than to push a particular viewpoint of my own.

I may sometimes appear to contradict myself. If so, it is probably be-cause I take different points of view in different chapters. I may also sometimes wander off the main point—publishing—and get into other aspects of media. This is all too easy to do. For example, where once the difference between a book and a motion picture was clear to all, it is becoming ever less so, as we move into a multimedia world and more and more books are coming out as multimedia productions.

Because I do not want to be a shill for the great, glorious electronic future does not mean that I don't think some of these bits of technology are great. But many have their side effects, and I have become increas-ingly leery of high-pressure salesmanship and of the technological imperative that says if I can conceive of it, it has to be built, and if it exists and is electronic, I have to own one.

As any number of people have pointed out before me, the rate of technological change today is greater than it has ever been. We do not have several generations to try out ideas before we incorporate them into our culture. Books have had a tremendous role in transmitting culture. Changing books too fast may have consequences we cannot foresee now. On the other hand, there was a world before books and if books as we know them today were to disappear, there will still be a world. Better? Worse? I don't know. *Neither does anyone else.* Make up your own mind.

A few words about what this book is *not*. It is not a how-to book and it is not a catalog of what is available electronically. The reason it is not a how-to book is that such books tend to emphasize mechanics and implicitly approve of what they teach. They may ignore effects or impact and that is my primary interest. The reason it is does not contain a catalog is that such a work would be obsolete the day it came off the press. It is also not a work of academic scholarship. I am trying to survey and explain issues, some of which are quite profound. But I cannot and make no attempt to summarize all the research on any one. I am content to bring issues to my readers' attention.

In an appendix are a number of statistical charts, put there to avoid too much technical detail in the main text. These may be browsed separately. Some, not all, are referred to in the text. In order to preserve a meaningful order among the charts, I have grouped them by subject, but the order of referral from the text is different from this order, hence references to figures numbered A1, A2, and so on are not in sequence by number.

Portions of the text have appeared in different form in the *Journal of Scholarly Publishing* and in a talk I gave to the Alumni Association of the School of Information and Library Science of the University of North Carolina. I am grateful for their permission to use that material.

For readers interested in pursuing any aspect of this subject, I have included some recommended additional readings, trying to keep away from overly technical or obscure material.

I acknowledge gratefully the help given me by a number of people. Albert Tedesco, my partner in a previous writing project, gave lavishly of criticism and suggestions. Others who read and commented on the manuscript were Mary Louise Meadow, Sandra Meadow, and Gretchen Whitney. I am indebted to several libraries and librarians at the University of Toronto where I have made use of the Faculty of Information Studies Library, the Bora Baskin Law Library, the Trinity College Library, and Robarts Library. And there was the Metro Toronto Reference Library. I got a great deal of help from and respect for the *Encyclopaedia Britannica* and The Dialog Corporation.

Chapter 1

Changing Media in a Changing World

In the beginning was the Word, and the Word was with God, and the Word was God.

The Gospel According to John 1:1

And further, by these, my son, be admonished;
of making many books there is no end;
and much study is a weariness of the flesh.
Let us hear the conclusion of the whole matter.

Ecclesiastes 12:12

I tell you, monsieur, the world's at an end. Never were there such breakings-out of the scholars! It's the accursed inventions of the age that are ruining everything — the artillery — the serpentines — the bombards — and, above all, the printing press, that German

pest! No more manuscripts — no more books! Printing puts an end
to bookselling — the end of the world is coming.

Victor Hugo, *Notre Dame de Paris*, 1883

Put writing in your heart that you may protect yourself from hard
labor of any kind and be a magistrate of high repute. The scribe is
released from manual tasks; it is he who commands.

From an Egyptian papyrus of about 1600 BCE[1]

The whole of existence is contained in words.... Words will never
die, so long as there are human beings to receive them. All our
knowledge and feelings can thereby be retrieved.

Meier Zable, quoted by Arnold Zable in *Jewels and Ashes*[2]

Wherever they burn books they will also, in the end, burn human
beings.

Heinrich Heine, *Almansor: A Tragedy*, 1823

The Power of the Written Word

These passages attest to the power of the written word. When most of
them were penned, what was really meant was the content of books, not
the objects themselves — except for Hugo's bookseller who foresaw in
printed books a loss of sales for the old-style ones. Burning a stack of
paper is not so bad. But when that stack contains words that inspire, the
burning is seen to be not of the paper but of the words, ideas, collective
history, and beliefs they represent. And, as Heine said, those who would
destroy ideas would not hesitate to destroy the human being who held
them. Unfortunately, people do not always bother to make the distinction
between the idea and its recording, and in an age when there was only

one practical form of long distance transmission of ideas, the message and the medium were the same. Today, there are more ways to communicate. Books are no longer the sole repositories of knowledge. But they retain an emotional hold on us. Today, also, we face the possibility of rapid and radical change in what those things we call books look like and feel like and in their contextual structure. Some of us eagerly embrace the changes. Some doggedly resist them.

Whatever side you may take, the prospect of change in books is important. They have played too great a role, not only in our lives but in our whole civilization, for us to be casual about seeing them change significantly. Unless, of course, the change is for the better.

In the last quarter century there have been, or seem to have been, more changes in media and telecommunications than in all history before that. We have seen cable television become the dominant transmission mode for that medium; seen computers enter most offices and a large percentage of homes; seen compact discs adapted to computers and able to hold a full encyclopedia or hundreds of novels; seen a marvel called the Internet grab our imaginations, time, and dollars; and seen wireless telephone and now wireless TV signals covering a whole continent. And changes continue to come at an almost frightening rate. What does this mean? What *effect* will these changes have on each of us? Are they real and significant or just the vacuous hype of technology companies? Will they affect us, primarily, or the next generations? And, if the next generations, will those people feel the same way about them that we do? In other words, are we bothered by the nature of the changes, or by changes per se?

In the past, some major social changes have been associated with media technology, although it is not always clear whether the media caused the social change or the social situation led to adoption of an invention that might otherwise have gone unused. Did the alphabet make Westerners into logical, scientific thinkers, or did that mode of thinking lead to the invention of the alphabet? Did printing cause the renaissance and enlightenment in Europe, or did these developments cause greater use of printing? We could not have had the suburbs and skyscrapers we now enjoy (if enjoy is the right word) without the telephone, automobile, and elevator. But did the desire for this kind of growth spur the inventions, or did the inventions actualize a latent desire? Possibly, it is not a matter of A causing B, but of A and B — invention and social change — interacting, or even of condition or event C causing both A and B. The important point to recognize is the frequency with which major media changes and social changes occur together.

We are going to be looking primarily at the world of publishing and how it is or might be affected by all this electronic technology or, more importantly, how we or our children might be affected. But when we look at publishing, we cannot limit ourselves to books, newspapers, or magazines. We have to consider mail, broadcasting, telephone, and motion pictures as well, because all these media are being thrust into a giant melting pot and when they come out, they may be one uniform medium or there may be entirely new media we have not considered before.

Lately, we are hearing the term *content* used almost as if it were a new technological invention. It is not. Content is what it is all about. Content is the story, the instructional text, the theatrical performance. The media are there to deliver it.

A question I am frequently asked, as soon as someone learns of my interest in electronic publishing, is, "Will the printed book disappear?" I think this is the wrong question. I know of no one who is actively trying to make it disappear. What is happening is that technologies other than print have been developed and they may challenge print for some types of books. The meaningful questions are more like, "What are the new media? How will we use them? In what ways are they better than print, and in what ways worse?"

The idea of what a book is may change radically. Newspapers may continue to be published, but not necessarily on paper. Television may become more interactive, giving us not only control over selection of the content (program) we wish to view, but even over the content itself. We may be able to filter out violence, see news only about our favorite sports teams, see everything that happened to our favorite politician yesterday, tell the newscaster what we want to hear, or even tell a director how we want a story to end. Books and movies might merge into a single medium and all novels might be acted out, with scenery, background music, and sound effects.

What would such changes do to our educational system, to our children's ability to read, or their *need* to read? Will they affect the very way we think? Will we have to learn to read all over again to keep up with the new kinds of media? I have no delusions about being able to answer all these questions. My objective is to raise them, to try to get beyond the hype of the new technology and to invite thinking people to consider the issues. Let's briefly review some of these issues before going deeper into them.

The Medium and the Message

Marshall McLuhan is probably best remembered for his aphorism *the medium is the message*.[3] While interpreting McLuhan is always hazardous, he did, in fact, like aphorisms such as this, and he felt that his mission was to make people think, not to answer all questions. Clearly, the medium by which a message is conveyed affects the manner in which the recipient understands it. Seeing a football game is much different from reading about it. Seeing a war through live color television, as we did those in Vietnam or the Persian Gulf, is far different from reading about it or seeing still pictures in a news magazine. Many observers felt the O.J. Simpson trial was played to the television audience, not the jury. Mentally merging the medium and the message (content) is fairly common. We have probably all said something like. "I heard it on the radio" or "I read it in the newspaper" to mean that, because of the medium, the content is assumed unimpeachable.

But we cannot take the aphorism as literal and universal. The medium *conveys* the message and it affects the way in which the recipient perceives it. It does not *make* or *constitute* the message. Television may portray excessively violent fictional scenes, but the electronic machinery of TV does not write the scripts or literally kill people. It may inspire killing. On the other hand, we must admit that the chance to get on TV may have created a great many news stories about events that otherwise might not have happened. A friend of mine who was at the University of California at Berkeley in the early 1960s, when student demonstrations, which peaked in the late 60s, were just beginning, reported seeing one such incident. A rather small group of students were casually loitering around a campus gate. A mobile TV news crew arrived. Almost immediately, the group became a crowd, and the casualness became high intensity. After a few minutes of camera work, the TV crew departed and the crowd reverted to its previous lethargic state. Still, we have to be careful to distinguish between the crowd's behavior being caused by the television mechanism and by the crowd's desire for publicity. In other words, there is a difference between the medium and the message. But in this case the very presence of a real-time (shown as it happens) pictorial medium led to the creation of the content — the student actions.

Computers are now in about one-third of North American homes and almost every office, factory, and laboratory. Telephones and televisions are in well over 90 percent of U.S. and Canadian homes. VCRs are found in about two-thirds of homes in the two countries. Many people seem at least somewhat concerned about the effect of these electronic media on their children, although possibly for different reasons. Some are bothered by the explicit sex and violence, some by the quality of televised fiction compared with what is available in print form. Some resent the trivialization of non-fiction, reducing electioneering to 30-second spot announcements or rehearsed debates. Some really hate the general attraction TV and computer games have for children, luring them away not only from reading but from physical exercise. Just like the early Berkeley demonstrations, television caters to a craving for instant gratification, something hard to achieve through the printed word.

There is another side, or various other sides, to this issue. New media, in general, are distrusted by many people. I suspect that most of those who resent TV as inherently without merit and mind-destroying are too young to recall (as I do) parental disapproval of comic books in the 1930s or 1940s. Possibly they had a negative effect on us. How can anyone know? Too many other things in the world changed for us to be able to put the blame on Superman. We are different people than our parents were; we grew up in a different world. Before comic books there were the dime novels, which I didn't experience, but which surely were seen as threats to civilization in the homes of the cultured of their day.

There is or can be a positive side of television. TV news has generally been granted the capability to make world events more real, as was especially demonstrated during the Vietnam war or the first moon landing. There is good quality drama on TV. There are excellent children's shows, such as *Sesame Street* or *Mr. Rogers' Neighborhood*. We have been hearing for years about TV's potential for education which, as yet unfulfilled, may still be realized.

Electronic Mail

Computers can now be linked together through telephone lines. This has opened up the capability of communicating through electronic mail, or e-mail, with anyone, anywhere, who also has a phone-linked computer. Soon, we will drop the copper umbilical cord and be able to e-mail from home, car, cottage, or canoe. Another form of "mail" is the electronic bulletin board on which anyone can post a message for all to read, and which can be responded to publicly or privately by any reader. The number of e-mail users is unknown but is believed to be in the tens of millions and still growing. These services can be used by researchers to find solutions to difficult problems, by cooks looking for a special recipe, gardeners looking for weed-control advice, or lonely people seeking dates or pen pals. Researchers have found that users of such services tend to be more open and more expressive of their feelings than users of the telephone or written mail.[4] Thus, computer-mediated, electronic mail does not always have the effect of dehumanizing its users as many predicted it would. To the contrary, for some people who have difficulty with more traditional forms of human interaction, e-mail provides a welcome opportunity for self-expression that they may feel is denied by our more conventional communications media.

In my own case, I moved in 1982 from the East to the West Coast while in the process of coauthoring a book. My writing partner and I relied on e-mail and were able to exchange drafts and comments easily and quickly. In my western job, I moved into an office that was already heavily involved with e-mail. One of my jobs was communicating with field offices across the United States and Canada and in Europe and Australia. E-mail was a great solution to time zone problems. You always could count on getting a reply within 24 hours. I also found myself and others being more informal than in traditional business letters and one result was that when I met some of my correspondents for the first time, face to face, it was like meeting old friends. Anything but a dehumanizing experience.

Electronic mail today is not a means of publication as we have generally come to think of it. But the technology that delivers it, typified by the Internet, could become a distribution channel for any kind of digital information or form of published work: newspapers, journals, books, radio programs, or movies. The genesis of scientific journals[5] was letters circulated privately among scientists. Much would have to change for this new form to take over completely, but change, after all, is what this book is all about.

Multimedia

Recently, multimedia capability has appeared on the computer scene. These machines have color screens capable of displaying high-resolution pictorial images in addition to text and numbers. They produce stereophonic audio output. They are driven either by a compact disc or information received from another, remote computer. The compact disc, more precisely Compact Disc Read-Only Memory (CD-ROM), is basically the same device used in stereo music systems, the replacement for the phonograph record. It is compact in the sense that a huge amount of information can be recorded in a small area and it is read-only in the sense that, once recorded, these discs cannot be written on, that is, cannot have their content altered. They are somewhat like phonograph records. Making the master may be expensive, but after that, large numbers of copies are quite cheap per unit.

The read-only aspect of compact discs is changing, and we are now seeing compact discs that can be written on by the user's computer. The compactness means that 500-600 million characters can be written on such a disc today, enough for the full text of hundreds of books. Once again, all sorts of magnificent accomplishments have been claimed or foretold for computers using these discs, largely but not totally unrealized. Digital video discs with some ten to twenty times the capacity of CD-ROMs are now on the market. This means thousands of present-day books or a full-length motion picture recorded on a single 4-inch diameter disc.

What does this mean for publishing? The text of large numbers of interrelated or cross-referenced books can already be recorded on a single disc. The creation and editing costs may rise considerably. But the cost of production, after the editing and markup necessary to any good quality publication, will go down significantly. The cost of shipping and warehousing books can go down enormously. Partially reducing these savings, the inclusion of background music and full-motion video in a "book" will increase its composition cost considerably. In today's technology, we really do not have a computer the right size and weight to make reading books on disc anywhere near as convenient as reading traditional ones. But that will change. We'll go into the economics of publishing later. Today, the biggest barriers to an all-disc publishing world are the lack of the reading device and the reading public's devotion to the form of book it grew up with.

Airwaves, Cables, Telephones, and Death Stars

The technological innovations in computing and multimedia computing alone might have been enough to make a revolution in publishing, entertainment, and education. But still more is happening.

Television was initially limited to 12 VHF channels, then expanded to 60 by use of UHF. No community I know of ever used all of these, but in a major market such as New York City there would be a great many in use, while in sparsely populated, far northern Ontario, the most one could expect to receive is one or two "over the air" channels. Cable offers the opportunity to about double this capacity although, again it is rare to find all hundred or so channels in actual use. But we creep farther up every year on the couch potato's utopia of 100-500 active channels.

By the way, couch potatoes: I had one experience of living in a house with a satellite TV system. It could gather signals from five different satellites, a total of about 100 channels, many (the principal American networks) duplicated on all services. It used to take me 5-10 minutes of browsing in the program listings just to find who was carrying a given baseball game. The more offerings there are, the greater the need for information about the offerings in the form of guides or indexes. The "invention" of *TV Guide* was a publishing master stroke. It will need to be computer based in the future and interactive, allowing the user to browse easily, not just to watch a listing slowly scrolling by, because there will be so much to look through.

Cable TV was once called Community Antenna TV and was originally developed to overcome certain natural or human-made obstacles to a broadcast signal, such as mountains or skyscrapers. Today, it tends to give better-quality signals and offers far more channels than are broadcast over the air in any community. The disadvantage of cable is, well, the cable. To broadcast TV, one erects a tower and sends out electromagnetic waves that, at their frequency, tend to have a range of around 80 kilometers, or 50 miles. But they can reach just about everyone within this radius not blocked by a mountain or building. To send signals over a cable, it is necessary to lay cable from the originator, or "head end," to every building that houses recipients — a very expensive undertaking. But these cables are capable of carrying a great deal of information, far more than the telephone line into your house can carry.

The cable people, not unexpectedly, are looking for uses of their expensive cables other than just carrying conventional TV signals. One possibility is sending recorded motion pictures to an individual subscriber, what is called movies or video on demand. The customer tells the studio what show is wanted and a compacted copy is sent by cable for a one-time showing in the home. Why make the long trek to the video store?

(That's a rhetorical question. There are reasons pro and con for trek-king.) It is also possible to do a much better job of TV shopping than is now generally done, or to do banking. Another use for the TV cable is to handle the increasing amount of traffic that computers are sending over the telephone lines, which were not designed for this level of use. By the way, computer-to-computer calls tend to last much longer than human-to-human voice calls, thus tying up telephone equipment for much longer. This situation has been the rationale claimed for telephone company re-quests for increases in local telephone service fees.

Not to be outdone, the telephone people have worked out ways to compact a message, making it possible, for example, to transmit movies over the telephone lines, thereby possibly competing with the cable TV operators. Telephone companies are also trying to get more and more involved in banking-by-phone and buying-by-phone. All sorts of call-forwarding, voice mail (recording of messages), and caller identification services are aimed at increasing telephone usage and, of course, telephone company revenues.

Also not to be outdone, broadcasters such as HBO are sending out ever more scheduled motion pictures. Enough variety in this mode may impact the market for on-demand services. Little is sure in this business.

There have been experiments with putting TV transmitters in airplanes as far back as the 1940s. These would broadcast educational programs to a limited region, whether a single city or a broad Arctic area.[6] Communi-cation satellites were first put up in the 1970s. The modern versions have, among other possibilities, the capability to broadcast a TV signal to a very wide region, or "footprint," on the earth's surface. The satellites are not exactly cheap, but the broadcasters do not have to wire a whole coun-try in order to use one. They do have to entice their customers to buy receiving antennas. Until very recently, these were the rather large (about 10-15 feet in diameter) "dishes" often seen on or near rural homes and urban bars. Now, we can have "direct broadcast satellite" (DBS) or "di-rect-to-home television" via a small antenna dish *inside* the house, more like 18 *inches* in diameter. This form of broadcasting is in direct, head-to-head competition with cable TV and has been referred to as the "death star," death, that is, potentially to cable operators.

During all this time, broadcasting over the airwaves has continued, but engenders relatively little excitement. There is high-definition television (HDTV), a largely Japanese-promoted development that could give us signals with higher resolution (in effect, less fuzziness) than conventional

TV. But this is hung up now in battles involving national pride, issues of compatibility, and possible other concepts of what better broadcasting means.

A couple of other recent innovations: Radio programs have been recorded digitally (as we now record music), then compacted and transmitted via telephone to a recipient's computer, a form of e-mail. On receipt, it is reconstructed and "broadcast" or "narrowcast" through a multimedia computer. Television programs can be similarly handled, although they require considerably more storage and transmission capacity. CNN, the Atlanta-based, international TV news broadcaster, in partnership with Intel, the computer chip people, introduced *CNN at Work*, a means of feeding the CNN cable-delivered news programs into a recipient's network of computers. It is intended for office use (large computers) where the broadcast or any part of it can be sent to all computers connected to a main one, and the programs can, if desired, be stored for future reference. It can be used to provide employees of a company with a custom-edited news program about their industry.

This technology is still rather expensive for home use today. For example, most people probably have much smaller screens on their computers than on their TVs. The two machines are in different rooms, making channel hopping awkward. But it could easily evolve into a new means of broadcasting, using TV cables as channels and a computer display instead of the television receiver. Or, as we pointed out earlier, if the Internet were used, the benefit might accrue to telephone companies. As we shall see in a later chapter, the Internet is growing phenomenally, so its custom would be quite a bonanza for any telecommunications carrier.

What we have is a situation that is almost frantic. In fact, if you are an investor in mass telecommunication, it *is* frantic. What were once carefully and legally separated media, telephone and television — are beginning to compete, each in the other's former controlled monopoly market. Computers are dramatically increasing the amount of traffic going over telephone lines, generating enough revenue to make that competition ever more heated. As computers become multimedia devices, capable of having text (characters typed on a keyboard) as well as sound (music and the spoken word) and pictures input as well as output, the difference between a computer and a television begins to fade away. But because computers can be *interactive*, that is, able to react to an individual user's input, the computer can do more in the way of entertainment or instruction than the TV set can

Hypertext and Interactivity: Hyperactivity?

A new form of writing, called *hypertext* is achieving some popularity. Basically, it is a form in which the reader can hop around at will through the text, reading the pages of a story or textbook in any order chosen. The reader can, for example, choose to read all about one character before another, or about a place, or a relationship. Is this a "good" idea? Is it a good way to present instructional material? We really don't know yet. There is much to be said on both sides of the question.

A new term in computer science is *virtual reality* (VR). VR, as a concept, is not really new. Just the horrible name is new. As early as World War II there were machines that simulated an airplane cockpit and were used for pilot training. While these Link Trainers were not exactly like the real thing, they were close — cheaper, and far less dangerous than the real thing for training. Today, aircraft simulators, now far more sophisticated, are still used. Virtuality can bring this kind of simulated experience to many fields, making the user feel a real part of a scene that is actually only simulated, taking actions that seem real and seeing the consequences simulated but appearing real. You could simulate a human body and the process of performing surgery. You could test extravehicular equipment for use on space flights.

One of the most criticized aspects of television is that everything is fast-paced. There is lots of action. There is no time for reflection. If all or most books were to be published in hypertext, multimedia, or VR form, or any combination, there is a question of what that might do to the process of thinking and reflecting that we often associate with reading. Will the whole world become hyperactive?

Is the Rate More of a Problem than the Change?

Is it a good idea to change so fast? Will we or our children lose the ability to read? I don't mean literally lose the ability to recognize words. But as it becomes ever easier to search for and retrieve bits of text and illustrations from existing works, will they lose the ability to study and understand a difficult, complex text and add their own creative interpretations?

I believe that most, maybe even all, major changes in media were initially met with something between resistance and outright hostility. Not by everyone, of course. There are always early adopters of an innovation and it is likely that at least some of their motivation lies in being seen to be innovative. Others adopt early because the new technology is clearly a solution to a problem. For still others, though, the new technology means change, an uncertain future, the danger of losing status, money, even a career.

Consider Homer. He was believed to have been a story- and newsteller. In an oral culture, he would have had a prodigious memory, would have used the meter and rhyme of poetry to help him remember the stories, and would have varied his stories to meet local circumstances and audience receptivity. These variations would add to, not detract from, his attractiveness. Imagine, then, the reaction when the aged fellow grew infirm and wrote his stories out for a younger successor to use. So, one day the young man shows up in a town and *reads from a scroll*. The audiences would have been horrified. It would be like Pavarotti sending a young stand-in to appear on stage and play tapes. But there is always another side to these coins. As Jay Bolter pointed out, the written form was more sophisticated poetry.[7] Later peoples could appreciate the new form. Those present at the creation probably did not.

Early movies and television were not seen as art forms, at least not by many. Even the telephone was vilified as an instrument of the devil when it first appeared. Telegraph and radio came into public awareness gradually, first only used by professionals. Hence, while there was the usual feeling that this form of communication could not be done, once done they were accepted. The telegraph in its early days was largely a railroad tool and did not directly intrude into the lives of ordinary people. Eventually it became a message service that, for the average person, meant only congratulatory messages or bad news. Radio was initially wireless telegraphy that enabled telegrams to be sent to ships at sea.

Consider the Victor Hugo quotation at the beginning of this chapter. He put these words in the mouth of a character who was a bookseller at the University of Paris in the 15th century. The point is that this character would have no reason to be against books themselves, but he would rail against the way he anticipated printed ones would change his life. As so often happens, he was wrong. I'm sure that real-life, modern university bookstores make more money than this character from fiction would ever have dreamed of.

When innovators can control the introduction of the new medium, they can do it gradually and painlessly. When the long-playing phonograph record (33 1/3 rpm) was introduced in the late 1940s, along with high-fidelity phonograph equipment, owners of the older equipment and records

(78 rpm) faced a great cost to make the switch. The new equipment was much better, but expensive, and it was only worth it if the old records were replaced with the new higher-quality recordings. One ploy was to offer a large number of records free, or very cheaply, with the purchase of a new phonograph. And, of course, the new equipment could play the old records as well. It helped. If a customer invested in a machine, he or she became an instant collector of the new records, now with a vested interest in supporting the new medium.

It was a very long time from the invention of the alphabet to the point where most government or commercial leaders could read and write — a gradual change. It was a short time from the wide availability of television to the near saturation of the market, and one result is that it is still controversial.

The great challenge we face today is coping with rapid change in media. Will we be hopelessly left behind if we do not adopt the cellular telephone? Will our children be denied meaningful careers if they do not learn to use computers in second grade? Should we scrap our cable box and buy a direct TV receiver? Do we have to worry that next year John Grisham or Danielle Steel will be published only on discs? Will our own jobs be next to be taken over by computers? Say, it was one thing to eliminate some clerk-typist or bank teller jobs, but what about doctors and lawyers for heaven's sake? In 1976, Joseph Weizenbaum,[8] an MIT computer science professor, reported a conversation with a colleague in which the colleague asked, "What do judges know that we cannot tell a computer?" The colleague's own answer was "Nothing." I doubt anyone today who knows any computer science really believes that a computer could do the job of a courtroom judge, but do the people who hire and fire lawyers know that? Are we in danger that some cost-strapped government might try to use a computer in place of a judge?

Our children and grandchildren will not only live in a different world, they will to some extent be different kinds of people than we are. They will not be traumatized by the great changes in media and education we are now facing, because these changes will already have taken place when these children became adults.

I wish I could say, "Sit back, relax. I'll take you through all the issues and tell you what's going to happen and how to cope with the changes." I can't. What I hope to do is raise the issues and get you thinking about them.

Chapter 2

Media and Information

The word *media*, from a word originally meaning "middle," has come to be used often without reference to what the media are in the middle *of*. What they (I'm old-fashioned — it's a plural word) are in the middle of is communication. Radio, television, newspapers, and the post office are media that carry messages from a source to a destination. What is in these messages? Information, or potential information.

What Is Information?

Information is a word with a maddeningly large number of meanings. General dictionaries tend to define it in terms of such related words as *data*, *intelligence*, or *knowledge*. Among people who study information professionally (information scientists, librarians, and intelligence analysts, for example) there is a tendency at least to make a distinction between, let's call it *stuff*, that has not been read or evaluated, and that which has been looked at and found of use or interest or at least understood. *Information* usually gets the higher of these meanings — it represents some form of communicated message that has value or meaning. The raw stuff is often called *data*.[1]

Claude Shannon, a Bell Telephone Laboratories scientist, developed a very formal model of communication[1] as a six-part process, illustrated in Figure 1. A message originates when its *originator* selects some symbols to transmit. These go through a *transmitter,* which puts the message in a form to be transmitted through a *channel*. A *receiver* accepts the message from the channel, transforms it, and hands it off to the *destination*, or recipient. These transformations are necessary because the message is usually encoded to go through a channel, whether into Morse code, or drum beats, or sound waves. Noise is added to almost any transmission by such sources as electrical storms or video signals bounced off large buildings. This model can be applied at many levels: a person as originator, microphone as transmitter, microwaves as channel, a television set as receiver, and a person as ultimate destination. This is shown in Figure 2.

More elaborate sets of definitions make distinctions among data, information, news, intelligence, knowledge, and wisdom. Here are some definitions I have used myself.[2]

Data

A datum (singular) is a string of elementary symbols, such as integers or letters. These symbols are not always meaningful. The stock market tables in the newspapers do not mean much to me. I can read and understand an individual entry, but not the "message" of the complete day's activities. Which stocks advanced, which declined? What does the pattern imply for the future? The baseball scores in the same paper may not mean much to another. A paper written in Hindi will mean nothing to either of us, if we do not read this language.

Information

This word has no universally accepted meaning, but generally it carries the connotation of evaluated, validated, or useful data. Information, to most of us, has meaning. We don't tend to use that word when the symbols are meaningless.

The same Claude Shannon developed a mathematical theory of communication, including a definition of *information*,[3] essentially calling it a measure of the absence of uncertainty. If a message has meaning, say the

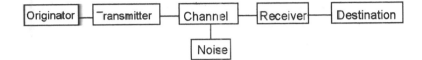

Figure 1. Shannon's model of communication. *In the general model, an* originator *decides what to send, does so through a* transmitter, *which sends the signal through a* channel *to a* receiver, *which in turn presents it to the final* destination. *That basic concept is repeated again and again in the communication between components of a larger system.* Noise *is anything that alters the signals as originally sent.*

latest price of General Motors stock, then it reduces uncertainty about that price or the implications of a change in price to the market in general. If you have no knowledge of or interest in baseball, the news that the Dodgers beat the Giants dispels no uncertainty — there was none to begin with; not that you already knew, but you didn't care. The statement "1 + 2 = 3," taken by itself, also has no information content for most of us because "everyone" already knows it. It brings nothing new or unexpected to us

A definition often used in physics or cybernetics is that information is *data that changes the state of a system that perceives it,* whether a computer or a brain If you pay no attention whatever to baseball, the scores do not change your mind or memory at all. A person who carefully listens to a sermon or political speech and becomes roused to action has had the state of his or her mind changed by information. A person who sleeps or daydreams through the speech has experienced no change. Both the daydreamer and the one aroused may have heard the same words — data to one, information or wisdom to the other.

News

News is usually defined similarly to *information.* It is a message, unexpected to some extent, that is believed true. If it is unexpected and believed, it changes the recipient. When a television news program delivers nothing really new or meaningful to the recipient, it is still called

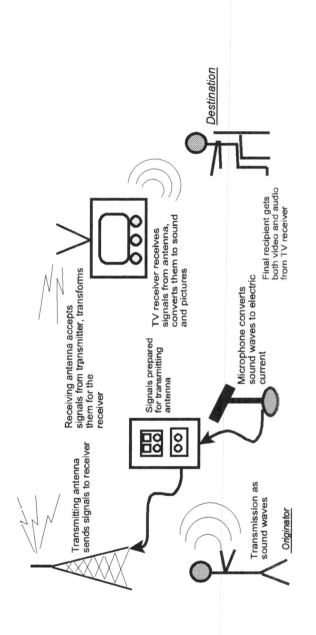

Figure 2. Shannon's model applied. *When a person speaks before a microphone and television camera, the mike and camera convert sound and images into electrical energy for reconversion into electromagnetic waves by a transmitter. The waves are received by a local viewer's antenna, converted to sound and images within the TV receiver, and presented as such to the ultimate destination.*

a news program, but it contains no news. People in Toronto have great fun at the expense of the Buffalo, New York, evening TV news because it seems to consist mainly of reports of fires in the nearby town of Cheektowaga, not of great significance if you are not in Buffalo, and perhaps no even if you are.

Knowledge

In general usage, knowledge seems to represent a higher degree of certainty or validity than information. My desktop dictionary suggests that information is a set of facts, while knowledge connotes understanding, with the implication that not all information is necessarily understood. To have understanding of a datum, it must be presented in a context, a background of previous information on the part of the recipient that came from previously accumulated and assimilated information. I think of knowledge, then, as an accumulation of integrated information in a person, group, or even a machine.

It is possible to insist that data and information are the same, but hardly anyone would equate data with knowledge. Sometimes, *information* is used just as I have used *data*. Daniel Boorstin is among the distinguished writers who have used it that way. He expressed his view of the difference between knowledge and information this way: "While knowledge is orderly & cumulative, information is random & miscellaneous. We are flooded by messages from the instant — everywhere in excruciating profusion."[4] What is important is to recognize that there is a difference. The names we give are less important than this reality.

Meaning

This is the most difficult of all the descriptors of information to define. "[N]o word ever has exactly the same meaning twice," said the semanticist Samuel Hayakawa.[5] Or, as Colin Cherry put it:

> The full meaning of a word does not appear until it is placed in its context, and the context may serve an extremely subtle function — as with puns, or *double entendre*. And even then the "meaning" will depend upon the listener, upon the speaker, upon their entire experience of the language, upon their knowledge of one another, and upon the whole situation.[6]

Wisdom

A person who has wisdom does not necessarily have more data or facts than others, but the wise person's utterances are more likely to be accepted by a community of "users" and to provide insight to matters of importance. Insight may be recognition of new relationships among observations not previously recognized as related. Our own political and religious leaders are wise; the nuts from the other factions are not.

What all this means is that most of us, when we are careful, recognize these gradations or something close to them. When we are not careful, we tend to call it all information and we lose the distinctions among messages or communications that are true, important, or wise and those that are merely strings of symbols. There is no end of declarations that we are in the midst of an information explosion or that we have an information overload. But such statements imply the data meaning. True information ("true" meaning my definition of it) requires that a recipient understand or at least be changed by a stream of data. If more messages arrive or are available than one can assimilate, that which is not taken in and digested is not information! The term *information* implies that understanding has occurred or a conscious choice has been made. A stream of data is just a stream of data. That is an extremely important distinction.

Why does it all matter? We're about to start considering media, and media are the physical means of conveying ... data, information, and wisdom.

What Is a Medium?

Medium is almost as bad as *information* in its number of different meanings and the ease with which we can become confused among them. A medium is:

The physical means of recording information or entering it into a transmission system: a pen, a typewriter, a telephone, a printing press, a microphone; or

A person who receives or perceives information from a source and in some form translates it to a different form for the recipient, as a spiritual medium, a painter, a poet, a news reporter; or

The material on which symbols conveying information are recorded: paper, photographic film, canvas, videotape; or

The means for transmitting information-bearing symbols: telephone wires or cables, electromagnetic waves used by radio and television, the post office carrying printed mail; or

An institution or organization that collectively produces messages to be recorded or transmitted: the University of Chicago, Public School No. 5 in Paterson, N.J., the New York Times Co., CNN, Walt Disney Studios.

The distinctions become important. The physical medium, as McLuhan said, affects the interpretation of content. The institutional medium, say a news reporting or motion picture-making organization, *makes* the content. They compose and tell the story. They report the news, whether or not the incidents they report ever happened. The medium as communications carrier conveys the information that someone has collected and prepared for distribution. As noted earlier, a television news program is a framework, a package for containing news. It may lack the tangibility of a newspaper, but it is like a newspaper in that its physical existence does not guarantee its content. There may be no real news. The paper will still be published, the TV anchor will still speak some words.

Telephone companies are considered to be common carriers, meaning that their services are open to all, and on the same basis. There can be discounts for volume business, but everyone with the same volume is supposed to qualify for the same discount. There cannot be one price for friends of corporate officers, another price for all others. Broadcasting companies are not common carriers. They carry the messages they want to send; they do not accept whatever the public asks them to broadcast. Yes, the government imposes some demands or limits on what stations must or must not broadcast, but basically the choice is up to the carrier. You or I cannot deliver a script to a TV broadcaster and demand that it be performed.

Without meaning to defend various media that are often criticized for their content it is important to recognize the difference between the medium as technology and as the organization that creates or selects messages. If the messages are unpopular with some segments of the public, who or what is to blame? Television, an inanimate technology, often receives broad criticism for content that should be aimed at the producers. Common carriers, because they do not control content, are rarely criticized for the messages they carry. Recently, the Internet has come under considerable fire in several countries because of its hate or pornographic content. The Internet is not a common carrier. It is subject, at present, to no licensing or regulatory authority. Printing technology and the book publishing industry as a whole almost never suffer abuse intended for authors. Books may be seized or suppressed by overeager customs officials, but only the most heinous of governments will suppress *publishing*. But there is certainly no lack of denunciation of individual printed items. Don't shoot the messenger. If any shooting is necessary, aim at the authors or producers.

What Are the Important Media?

The media that are most used and most influential in human communication are: voice and language, writing and printing, graphic art including photography, telephone, radio, cinema, television, and the new computer-based communication forms, about which I'll say more later. There are other media: plastic art (three-dimensional art or sculpture, pottery, etc.), architecture, music, and dance, but these all share the characteristic that "messages" in the medium are not created by as many people as use, say, oral language.

Even in the first listed group, some media are variants of others, not fundamentally different, as writing derived from graphic art. Television, as a technology, is a variation on radio. In TV, not only sound but also visual images are converted from their original form and made into electromagnetic waves for transmission and eventual reconversion by the receiver. But TV has had such a different impact that we tend to treat it as a different medium. Similarly, cinema is an extension of photography and, as noted, writing is an extension of graphic art, but in both cases the new form has come to be so specialized that we treat it as separate.

Today, we have computers that communicate to and from humans in written, graphic, oral, and touch form and can communicate transformed versions of messages in any of these media to other computers, thence to other humans. There is really nothing fundamentally new about these modes of communication, but the combination is or seems so different from anything we have had before that we again tend to treat the new computers collectively as a new medium. As we begin to criticize this new medium, whether constructively or not, we have to be careful to keep straight what we are criticizing — technology or content. As an example, 3-D movies that came out in the 1950s always seemed to have scenes in which something was thrown toward the camera, hence apparently toward the audience. It's very effective in the short term. You want to duck. But after a while it gets tiresome. This is not a technical requirement of the medium, it is a content choice that substitutes a cheap thrill for really effective use of a new medium. It seems to turn some viewers against the medium per se.

Are the Media Merging?

The usual justification given for awarding local telephone monopolies in the 19th century was that it was necessary to have compatible equipment and procedures so that any subscriber could talk to any other in the local area. The assumption was that if there were multiple telephone companies, there would be multiple and incompatible ways of doing things. Certainly, there was no early move to combine telephone with telegraph transmission technology. Why did the regulatory agencies not specify what we call today interconnect standards and allow for free competition? Even today, competition is rare at the local level for service by wire. The companies fighting for a share of the long-distance market had until recently been content to let the original monopolists deal with the local service, while they competed for the more lucrative long-distance market and complained about the price of local service. Essentially the same goes for cable television. In any neighborhood, there is one cable company. If they compete, it is for the monopoly in the next neighborhood. Since both telephone and television are moving toward wireless transmission, even these local monopolies will probably soon disappear.

When the computer industry began to flower in the 1950s there was no standard data recording medium that could be recorded by one type of computer and read by another, except punched cards, which were already becoming inadequate for the size of files the computers then used and the speed at which they were expected to run. In the early 1960s, for example, IBM had several families of computers on the market, all able to write on and read from the same type of magnetic tape, but each computer type might write in a different manner. An IBM 704 computer could not always read a tape made by an IBM 705. Gradually, the industry cleared up most of the computer interconnect problems, but still today you cannot always read a WordPerfect file made by a Macintosh, if you are using an IBM computer. The computer industry has successfully established standards for devices to be attached to a computer, such as printers, modems, or displays, but has not achieved this degree of standardization in the format of data to be stored or transmitted.

Now the electronics industry, or its avant-garde, is beginning to recognize that because telephone, radio, phonograph, television, facsimile, and computers all communicate by means of electronically transmitted messages, and because increasingly these messages are being sent in digital form, a whole new level of interchange standard is possible. As Nicholas Negroponte points out in *Being Digital* and Douglas Eisenhart in *Publishing in the Information Age*[7] all of the signals for electronic communication devices could be put in digital form and could carry with them instructions for interpretation or playback by the different media.

Today, facsimile uses voice telephone lines. When one fax machine calls another, they must exchange certain information about the form of messages and their respective capability to receive and interpret a message before the main message can be sent. If you have ever dialed a fax machine by accident and heard a high-pitched squeal instead of "Hello," what you heard was the start of the receiving machine telling you about itself — for example, what transmission speed it is capable of handling. If all electronic messages, regardless of origin or content, carried with them instructions telling what they contained and how they were structured, the receiver could decipher this information and use it to interpret the message as it wished or was able to do. The morning news might arrive in this generic form and be converted by your computer at home into print, into a format for display of text on a computer or TV screen, or be read to you by your computer, using a synthesized voice, male or female, with a New York or Alabama accent, as you chose. The pictures accompanying the news story could be played on TV as full-motion video or a few frames could be extracted and printed with your newspaper.

That this standardization of messages in digital form will actually come about is likely, but when this will happen is not a prediction I care to make. It makes some sense as a goal. It would leave it to the receiver, whether that means a person or machine, to decide how to read, see, hear, or maybe even feel, a message, and it would leave it to the person to decide what kinds of receiving equipment to have in the home or office.

You would not need a television receiver to receive video transmissions; they could be received and converted for display through your multimedia computer. Similarly, the publishing house would not have to decide in what form a story should be released. It could be in generic form, for use by any receiver with interpretive capability.

It would take a massive industry restructuring to reach this point. What difference would it all make? Once the world adopted such a standard and everyone had the requisite hardware and software, telecommunication and publishing would be much less expensive than now. It could also be far more varied because users would get to decide how they want to perceive movies, news, or novels. Authors or editors would have to consider this when assembling any work. It could also make for more imposed artistic uniformity in the interest of economy. Can a story, news or fiction, be written in so generic a style as to be suited for all presentation media? That is where I have some doubts about the brave new all-digital world.

Chapter 3
Some Media History

It is said with some wisdom that history is written by the winners. The history of media is also largely the history of media winners. Who are the losers? They are hardly known, so it is difficult to compare them with winners. What do we even mean by "winning"? In baseball, the winner is the team that scores the most runs. In popular music a winner is a song, or the performance of one, that sells well. It may be forgotten in a year. In media, a winner seems to be one that is adopted or used by a large number of people, one that is versatile in use, and one that lasts for years or even centuries or millennia.

Language and writing are media inventions that caught on and have lasted. They are not just widely used, they are almost universal. We do not, of course, all speak the same language but we all use a spoken language for communication among people who are physically close to each other. When people cannot speak or hear, they use sign language or learn to recognize words by the motion of the speaker's lips. When they cannot see, they use their fingers to feel the shape of embossed letters. But it is all language. Although writing is not as widely used as spoken language, its popularity and importance are still increasing. Increasing literacy is a major goal of most developing countries and of deprived segments of developed countries. Modern theorists of linguistics, such as Noam Chomsky and Steven Pinker, say that the propensity for language is inherent or instinctive in human beings.[1] We may have to be taught our own particular language, but the concept of language seems to come naturally. For children, then, learning to use the medium of spoken language is not particularly difficult and can be fun. Ease of learning and use are key factors in the decision to adopt any medium.

Writing, in the sense of penmanship, is not nearly as easy to learn as spoken language. I remember my own children, whom I immodestly felt to be very bright and to show it at early ages, seeming to hit a brick wall in school when they had to learn to write. It was slow, boring, and difficult compared with other forms of learning in grades 1 and 2. The only payoff, at least until later in life, was adult approval that came from being able to write. A similar feeling was expressed by a 12th-century scribe:

> If you do not know what writing is, you may think it is not especially difficult... Let me tell you that it is an arduous task: it destroys your eyesight, bends your spine, squeezes your stomach and your sides, pinches your lower back and makes your whole body ache.... Like the sailor arriving at the port, so the writer rejoices on arriving at the last line.[2]

But writing is so useful and versatile that in the developed world those without the ability to read and write are at a severe disadvantage in terms of finding and holding jobs and general functioning in society — reading catalogs, road signs, or instructions on everything from taking medicine to job requirements to government tax regulations. So, we seem to have been willing to pay the price for this difficult technology.

Graphic and plastic arts have been important parts of most civilizations. Since often the artworks have been almost all that survived, we do not know how popular these media were in the sense of widespread use. We do know that, however many people may have had a hand in creating them, these media have been important aspects of most cultures.

Language, writing, and art are very ancient. The date when language was "invented" is unknown and, of course, it would not have been a matter of inventing it all at once. It would have evolved over a great many years. The origin of writing is generally dated as about 6,000-5,000 BCE. Graphic art started far earlier, variously estimated as 40,000-14,000 BCE. But the next really big step in media, in the western world, was the movable-type printing press. The technology, or a form of it, existed in China much earlier, but it did not make a direct impact in Europe. Who knows, though? Perhaps Marco Polo mentioned it to someone who mentioned it to someone who ... and thence to Johannes Gutenberg who produced the first western typeset book around 1455. A process this casual is said to have led to the invention of the tabulating or "IBM" card.[3]

Books existed before Gutenberg, if we take a book to be an assemblage of pages bound on one side. Scrolls had existed for centuries, but these could be very cumbersome as the texts contained got larger. The book gave the beginnings of "random access," the ability of a reader to go quickly to any part of a collection of information. Here is a description of scrolls as used for administrative records in medieval England:

> Every current record must travel with the court, though the King [Henry II, ca 1160] rode sixty miles a day. The mainspring of the Chancery, round which all else revolved, was the Great Roll of the Pipe; a new roll was started on the first day of each year, ... with a copy of the first public document engrossed that year; the second document was stitched to the bottom of the first, until after a few months the roll was so thick that it did in fact resemble a drainpipe. There was no index, naturally, and the only way to consult a letter in the middle was to unroll the whole thing on the floor....[4]

The Book

Although possibly now threatened by discs as a medium for recording the printed word, the book is still regarded by many as an ideal information recording and transfer medium. It has become small enough to be carried conveniently (well, most books). It is inexpensive, although "inexpensive" is a relative term. It can be written on by those who care to do so, with marginal notes or yellow highlighting. It can endure for a very long time if properly cared for. It can be read in all sorts of environmental conditions: on the subway, in bed, at the beach (computer displays are not very good in bright sun). On the other hand, discs are even smaller, might last as long (we really will not know for a few hundred years), can be much cheaper, and do not use paper. The price of paper has shown an alarming and unpredictable tendency to soar upward and dive downward at times. Some people worry that another technological change will render a library full of CD-ROMs unusable unless it archives the obsolete reading equipment along with the texts.

This is, in effect, what has happened several times in one lifetime in the music recording business, as we went from wax cylinders to 78 rpm disc recordings, to 33⅓ rpm, to open reel tape, to tape cassettes, to compact discs, and now to digital tapes and discs. Each major shift in recording technology obsoleted the previous machinery or media in the sense that while some can run the older media, not much of the older equipment survived after the new arrived. This *could* happen with texts on disc but it need not, provided that the industry or its regulators pay sufficient attention to standards and methods of conversion, so that even if technology were to change, old media could easily be converted to new. The electronics industry has not had an outstanding record on this so far. But then, what technological industry has?

What we think of today as the book developed in about the second century CE and represented what was for many an improvement over scrolls. In Western society, books in that day were mostly read aloud in religious assemblies, not by individuals. This may have been caused by the cultural repression of the dark ages. In *The Confessions of St. Augustine*,[5] published in the 4th century CE, the author tells of visiting Bishop (later Saint) Ambrose of Milan and finding him alone and reading silently. Was it the silence or the intensity of concentration that was worthy of note? Augustine does not make it clear but we are left feeling that silent reading was not very common. These pre-Gutenberg books were written as a more or less continuous stream of characters. There were no chapters, tables of contents, indexes, or even spaces between words. These artifacts were gradually added and they came in something of a spurt in the 12th century.[6]

Gutenberg's invention changed the way books were produced, one consequence of which was a great increase in the number produced. Figure A10 in the Appendix shows estimates from the *Encyclopaedia Britannica* and UNESCO of the number of book titles produced in Europe and later worldwide. Figure A7 shows more detail for the 20th century. The numbers are still rising. One reason why the increase shortly after the new press was invented is so remarkable is that Gutenberg could not simply call his type supplier and ask for a new font, nor could he call the union hiring hall and ask for additional experienced printers. He and his associates had to develop the industry themselves. Actually, Gutenberg had partners who funded his development work and it was these partners, Fust and Schoeffer, who made the new printing into a successful business.[7] While there was a book trade before Gutenberg, it was nothing like what he unleashed. But this one man or one company were not completely responsible for the rising popularity of books. His was the time of the European renaissance and reformation. The typeset book was an invention for the times. A century or two earlier no one might have cared. A century later and western culture might not have developed as it did.

Over the years, the book has continued to evolve. It now comes in a great range of sizes, bindings, and styles. It may be a compendium of figures, a text, a book largely consisting of pictures, or any combination of these content types. Illustration — a pictorial presentation related to the text, as contrasted with illumination of the first character on a page — came into use very quickly after movable type. The first illustrated book was dated 1461. Separately printed graphics predated Gutenberg. It is interesting how quickly the new medium for text merged with the graphic medium. It took less than a decade. This was the beginning of multimedia. Today, there are books that can make sounds as pages are turned and books that contain objects, not merely illustrations. To me, the outstanding example of the latter type is *Pat the Bunny,*[8] a children's book I think of as indispensable to any parent of a one- or two-year-old child. It has text but also embedded pictures and such objects as a mirror and a piece of soft, fuzzy material in the shape of a bunny.

Still, when I talk about the prospect of electronic books, I find many, perhaps even most, book lovers react negatively. I say "still" because the book has already undergone so many changes, but many of us seem to feel we should now stop. We have reached perfection. But books will surely continue to change whether the direction is into electronics or not. Plastic is already sometimes seen as a cover or binding material and in a few cases as dust jacket material. Might we see it as the material of pages if the price of paper continues to rise? How does an all-plastic volume of Shakespearean sonnets strike you? How do you think it will strike your grandchildren?

The Telephone

The telephone was patented in the United States in 1876 by Alexander Graham Bell, a Scotsman who did his basic development work in Canada. While there was a sort of predecessor technology, the telegraph, it required a high degree of skill to operate. It was not for the average person. The telephone required some learning, but not much skill, and some experimenting with how it was to be used. At one time it was thought of as an entertainment medium and in Budapest in the 1890s, news, storytelling, poetry, and music were "broadcast" over the telephone on schedule. In the United States, the telephone was initially a point-to-point medium, meaning that there had to be a line — wire — from any caller to each destination that person wanted to communicate with. Bell recognized early on the need for switching centers, and he quickly developed them. Using this idea, each subscriber communicates only to the center, and the center connects the incoming line with the appropriate outgoing line. Without this development, the telephone could never have developed as it did, with almost every office or household in the industrial world connectable to almost every other one. We tend to take it for granted but it has to be one of the greatest technological feats in history that we can pick up a telephone in our homes and "dial" to any other telephone, anywhere else in the world.

What learning was involved? The early instruments would have been somewhat noisy, so a loud clear voice was necessary. The parties talking could not see each other so they had to identify themselves and just that small difference from face-to-face conversation would have taken some positive effort and must have felt awkward at first. Conversing with people you could not see was hardly a common occurrence before telephony. A hundred years later many people felt very self-conscious talking to an answering machine and at first many callers would simply hang up rather than do so. In only a few years, though, most people overcame the inhibition and adapted. Early telephone users also had to learn to so phrase their conversations that they could get or send secondary messages: "Did you hear and understand me?" "Yes, I did." These signals, in face-to-face conversation, tend to be sent by gesture and facial expression. This may seem a trivial adjustment to us, today. But it was not necessarily so for the people of the late 19th century. Peter Drucker reported on a management conference proposed in 1882 by the German post office to discuss how not to fear the telephone. "Nobody showed up. Invitees were insulted.... [T]hat they should use the telephone was unthinkable. The telephone was for underlings."[9] Is it possible that users of plastic or digital electronic books of the future will wonder why it was so hard for us to adjust to these, which seem to them quite ordinary objects?

In the development of a telephone industry, suppliers of the various components and related services were not already established and were clearly not listed in the Yellow Pages. Adequate cables, offering protection against weather, birds, and squirrels, had to be invented and manufactured. Installers, service people, and operators had to be trained. The instruments had to be manufactured in quantity. The number of telephones installed in the United States went from zero in 1876 to 150 million in 1993.[10]

Today, the telephone is found in 93.3 percent of homes in the United States and 98.5 percent in Canada.[11] It has become a virtual essential, like running water and electricity. Clearly, like the book, this was a communications medium that met the needs of its day, unstated at the time because most people could not conceive of the as-yet-uninvented device. And it was, or became, easy and pleasurable to use. The spread of urban populations out to the suburbs and up into high-rise buildings depended on the telephone as well as the automobile and elevator. Without these three means of transportation and communication, our modern North American society could not exist. In fact, Mosco pointed out that there is now a class of welfare called telephone welfare, to provide government support for telephone service to those unable to afford it.[12]

Mr. Bell was among the founders of the American Telephone and Telegraph Company, destined to become for a time one of the world's largest corporations. Governments in the U.S. decided that telephone service should be a monopoly in each locality in order to assure compatibility of equipment and connections. But it was regulated at both state and federal levels. A similar structure grew up in Canada where, generally, each province had its own telephone company, occasionally more than one, and in one case a single company, Bell Canada, serving the two most populous provinces, Ontario and Québec. Elsewhere in the world, telephone service was generally provided by the post office, a government agency. The United Kingdom, Jamaica, and Thailand are some of the countries that have recently privatized their telephone service, which perhaps indicates a worldwide trend. Elsewhere, today, the usual rule is that it is still a government service. The newer cellular telephone technology is more often a private service.

Radio and Television

The scientific work underlying radio and television was done in the 19th century. James Clerk Maxwell in England discovered electromagnetic waves, and Heinrich Hertz in Germany carried out the first transmissions using them. Guglielmo Marconi, an Italian working in England, who is generally credited with foreseeing commercial use for the emerging technology, patented a transmission system in 1896, and conducted the first trans-Atlantic transmission in 1900.

That was prologue. Radio as a practicality is essentially a 20th-century phenomenon. R.A. Fessenden, a Canadian working in the United States, transmitted music and speech in 1906. In 1920, the Westinghouse Corporation established radio station KDKA in Pittsburgh, Pennsylvania, to produce the first regularly scheduled broadcast programs, including news. Technical and commercial developments were rapid after that.

Originally called wireless telegraphy, radio was first seen as important for communication to and between ships at sea. As the *Titanic* was sinking, its radio signals brought some rescue aid. More might have been possible with modern rescue services and marine telecommunications. David Sarnoff, later to become head of Radio Corporation of America, was then a radiotelegraph operator for Marconi Wireless Telegraph Company. He is said to have heard the distress calls but could do nothing from New York to help. (There is some dispute about Sarnoff's role,[13] but none that wireless was used by the *Titanic*.) While radio is now mainly used for voice and music transmission, it is also still used as a message carrier, i.e., for point-to-point transmissions, as in taxi or ambulance dispatching. But the broadcasts of regularly scheduled programs of music, news, live descriptions of sports events, and drama are what quickly came to dominate the medium.

News coverage of the burning in New Jersey of the German airship the *Hindenberg*, with its passengers aboard, was broadcast live in 1937. Three years later, reports of the bombing of London during World War II were broadcast, notably, for American audiences, by famed reporter Edward R. Murrow. These broadcasts gave listeners a sense of closeness to the real events, a sense not possible through newspapers. In 1938 Orson Welles's radio dramatization of H.G. Wells's book, *War of the Worlds*, depicting a Martian invasion of New Jersey, caused great panic among some listeners who were not used to such realism in radio fiction and who thought the invasion was real.

The adaptation of radio to voice and music was rapid. Today, talk radio is gaining ground in news and music. In talk radio, listeners take part in making up the content. Drama, or any form of acted-out entertainment, largely migrated to television over the years. This is a case of one medium directly pushing out another. But new models and audiences were created for the old. Automobile drivers are a major radio market. A curious segment is attendees at a live sports event. People watching a game often like to listen to it on the radio at the same time. One version of the game, direct observation, gives an overall view, a sense of excitement, a sense of personal involvement. The other gives expert commentary on what is happening on the field.

It is radio that transmits television signals over a long distance, so we could not have broadcast TV without broadcast radio. Some early experiments go back to the late 19th century, when Paul Nipkow, in Germany, developed a working television using a mechanical scanner. The scanner is the device that "sees" a scene as a two-dimensional picture and converts it into the series of one-dimensional lines that "paint" the picture on the receiving set's screen. Vladimir Zworykin, a Russian-born American called the father of television, patented the iconoscope, an electronic scanner, in 1928 and things went rapidly forward from there. Well, some people think of it as forward. A better camera, the orthicon, was developed in 1927 by Philo Farnsworth and in that same year AT&T produced the first long-distance telecast, of a speech by then–U.S. Secretary of Commerce Herbert Hoover. In 1936 the British Broadcasting Corporation began the first regularly scheduled TV broadcasts.

Television never passed through a phase of being pictorial telegraphy. There were a few non-broadcast uses as early as the 1940s for education or training, but mostly it was a broadcast medium from the start. Its content was mainly entertainment (variety, comedy, drama), sports, and news. Entertainment and sports still dominate. A few news events, however, have had great influence on public acceptance and interest. In 1954 the U.S. Senate hearings on charges brought against the U.S. Army by Senator Joseph McCarthy were televised, the first such event so broadcast. It ran on and on like a soap opera, but here were real people pitted against each other in deeply emotional debate. Earlier, Edward R. Murrow had denounced McCarthy in his TV news program, the first major public figure to do so, and he a TV journalist, not a politician. In 1994-95 the O.J. Simpson trial, telecast via satellite and cable, attracted similar audiences, a unique event in juridical history.

Early TV drama was usually live and resembled live theater. Sets looked like theater sets, not real life. Lines were occasionally flubbed. Gradually, producers moved more into the cinematic mode. Programs were taped, not shown live. This allowed for on-location shooting, for retakes in case of error. Scenery was far more elaborate and realistic. Most TV news, still at its best covering major events live, remains thin in its routine coverage compared with the largest newspapers. Partly this seems to be due to TV viewers preferring short presentations — or did we just become accustomed to what we were given by the TV producers?

The adaptation to television by both producers and audiences was rapid. It did not take viewers much time or effort to learn to use the medium. Prior to the invention of the VCR and remote control, TV use was quite passive. One turned it on, selected a channel and sound level, and sat down. Today, the remote control unit makes viewing much more active. A much criticized medium, television is nonetheless often used and increasingly relied upon for information

Radio and television did not directly displace the printed word, but they encroach upon it. During World War II, in the 1940s, the American magazine *Life* was probably the principal source of visual information about the war for the American public. *Life* still exists but it is nothing like its former self in content or market dominance. Now, TV magazines such as *60 Minutes* rule what used to be *Life*'s domain. When newspapers "die," evening papers tend to go first. For the benefit of younger readers, in multi-newspaper cities it was common for one or more to be published in the morning, perhaps in several editions from around midnight to noon. Other papers published from early afternoon into evening. Evening papers are (or were) likely to be read in the living room, now shared with a television set. Morning papers are likely to be read during breakfast in the kitchen, a room shared with a stove and a toaster, but not commonly a television set. This is one theory of why the evening papers die first. Another is that migration of population to suburbs made delivery more difficult and expensive.

If broadcast media are competing with books and magazines, it is more likely because people are more attracted by the medium than the content. For all the sex and violence on TV, you can find at least as much in books. But television demands less of us in terms of mental effort and its images may appear more real to some viewers. The lower mental effort is at the heart of many peoples' objections to TV. It has become a medium of fast action, short presentations, and little depth, but it is easy to relax into. Dalton Camp, a columnist in the Toronto *Star*, wrote the following in reaction to the speed at which false statements attributed to him were published:

> Facts move at the same speed as fiction and the shelf life of a ringing truth is no longer than that of a simple lie. One of the more irritating things about living along the information highway is that information, disinformation and misinformation all travel at the same speed.[14]

The TV "book" is very different from the printed word, but the radio "book" could well be someone reading from the printed one. In fact, in 1945, the mayor of New York, Fiorello La Guardia, made a great hit during a strike of newspaper delivery drivers by reading the "funnies" to children (of all ages) on the radio. This could not work on TV, which demands constant action, rather than the display of panels of pictures or just a voice reading.

Computers and Computer Networks

The electronic computer is just over 50 years old. The first truly electronic one, the ENIAC, was developed at the University of Pennsylvania in 1946. The early models filled a room and the air-conditioning equipment needed to cool them filled another room. With the early models you brought your data and programs to the machine, sometimes waiting hours until it was your turn to "run." Gradually, starting in the 1950s, we began to attach what were first teletype machines or electric typewriters that enabled the computer user to be somewhat remote from the computer — typically in the same building but not in the same room. Eventually, the teletypes and typewriters were replaced with a computer terminal. This was essentially an electric typewriter or a combination of a keyboard and an electronic display unit that could communicate via the telephone network, making the distance between computer and user ever less important.

The monolithic "mainframe" computers could be difficult for the average person to use. (Heavy-duty electronic machinery was fixed into a steel structure or frame. Hence the name. Today's smaller devices do not need this support.) In the 1970s the personal computer was born. At first, this development made little difference, except perhaps to the Apple Computer Company and Microsoft Corporation. But rather quickly the PC became popular at home and on the desks of truly huge numbers of professional and clerical workers. (Although "PC" has become a symbol for an IBM-type computer, as distinct from an Apple-type computer, I use the term generically — a computer designed for use by one person, hence that person's personal computer.) These machines made computer users out of large numbers of people, many of whom would never have considered using a typewriter but are content to sit for hours before a computer screen.

Originally developed primarily for military use, in performing ord-
nance calculations, the early computers were designed and used for solving mathematical problems. Gradually and hesitantly, they were applied
to more mundane problems such as preparing a payroll. Payroll calculations are relatively simple but there are a great many of them and the
logical process of deciding what tax rate to use, whether or not to deduct
union dues, how much sick leave is available, etc., can become overwhelming for a large company. Similarly, inventory control starts out as
simply adding the number of items received and subtracting the number
sold, but it grows into a system for monitoring the rate of sales and replacement and determination of the best time to reorder and which supplier to use. Imagine trying to schedule the construction of a large building or a 100,000-ton ship, involving delivery of a huge number of parts
or materials to a limited storage space, without the aid of a computer.

Computer lore is replete with legends about early predictions of such
persons as Thomas Watson, Sr., the man who made IBM, that the world
market for large computers could be expressed as a number with a single
digit. Or the story that the developers of UNIVAC, one of the earliest
commercial computers, gave their machine 1,000 words (12,000 bytes)
of main memory because they could not conceive why anyone would
need more. Today, computers are found in about 36 percent of the homes
of Americans and Canadians, and the total number is in the millions. We
who are users have become accustomed to whatever size memory we
could afford last year being inadequate for next year's software. Is there
an end to this growth? Along with the growth in number and capacity has
always come at least a compensating shrinkage in the physical size. The
price of a portable PC in 1997 is about what it was (unadjusted for infla-
tion) when I bought my first one in 1981, but the performance difference
is extraordinary. Yes, someday there will probably be an end to the market growth. I do not expect to see it. As an example of the rapid change
that characterizes this industry, my newspaper yesterday (as this is written) reported that Bill Gates, the founder of Microsoft, is now the world's
richest person. The next day came the report that Smith-Corona, once
one of the top producers of typewriters, had gone into bankruptcy.

Just a few years before the PC, the U.S. Department of Defense created a communications system that connected a number of large computers far distant from one another. A user of one of them could direct that
machine to dial up another computer, and the user could use the remote

machine as if he or she were at the remote location. Without this capability, it could be prohibitively expensive to move a program from one machine to another because doing so might require extensive changes to the
software. But it was also necessary to economize on the use of connecting telephone lines or this whole undertaking would have been too expensive. This was done by taking message fragments, called packets, and
transmitting each one as a separate message, charging only for the transit
time of packets, not for the time between packets. It was as if you would
not be charged for that portion of a long-distance voice telephone call
that you spent thinking what to say or pausing between words. Someone
else's words would go out while you were pausing, and then the costs
could be shared. In computer communication, it is not two, but many that
share the same lines. This early computer network was called the
ARPANet, for the Advanced Research Projects Agency of the DoD.[15]

Before long, ARPANet technology went commercial and we had public networks that could be used by anyone — well, anyone with money
and know-how. For example, a subsidiary of Lockheed Aircraft Company, called Dialog, had developed a database service that would enable
a user to search a number of files to find what had been published in
fields such as education, chemistry, or engineering. This service was invaluable for research people, but it typically took about 15 minutes to do
a search. How many people, or even libraries, in the 1970s could afford
a 15-minute prime-time telephone call from the East Coast to the West
Coast and then pay for the computer time on top of that, and for the use
of the files on top of that? Because of the network design, the use of the
telephone lines was shared among a number of simultaneous users, bringing the price down to an affordable level. The cost of looking up a subject in Dialog's files was, and pretty much still is, about the price of an
average restaurant dinner. As I told many of my students at the time:
think about giving up that dinner in order to save hours of time in the
library and to be sure you get everything you need for that term paper.

As the 1980s came along, more and more networks were being established and more and more services were being offered. A popular one
was electronic mail. E-mail is something like old-fashioned telegraph
service, but with a difference. You type your own message into your computer. Zap. It is transmitted to the computer of your addressee, or to a
nearby computer and held until your correspondent turns on his or her
computer and asks for any incoming mail. If both of you are "on" your

machines at the same time, it is possible to transmit and receive as you type, so you are having a true, interactive conversation, but in typed, not spoken, form. In 1979 Compuserve made an e-mail service available to anyone with a PC and the device needed to connect it to the telephone system, a device called a modem. The cost was variable, but I paid about $3 an hour. And you could prepare a letter in advance, using your computer's word processor, then call Compuserve and send the message, bringing the cash price of sending a letter down to the neighborhood of that of a postage stamp.

Interactions of Media

To date, new media do not seem to have replaced the old completely. Storytelling, orally and in person, is still with us as an art form, although clearly not as important a medium as it must have been in pre-writing days or in modern oral cultures. Books, if we use a broadly generic meaning, to include scrolls, clay tablets, and the like, have not disappeared as radio, then television, then multimedia computers have come along. Television changed radio but did not eliminate it. The Internet may have impacted TV viewing, but we do not yet have data to show how, or how much. But, clearly, new media encroach on each other to some extent. Conversation is not the flourishing art it once was before radio and TV, but it has not disappeared. Many newspapers have died, but new ones come along each year. The ones that die seem to be largely general purpose and aimed at a wide audience. The new ones seem to be very specialized, either in terms of geographic coverage or content — neighborhood news, computer news, automobiles for sale.

A few media have virtually disappeared: the telegraph, carrier pigeon, vaudeville. The first two were replaced by newer technology that was faster, cheaper, or more reliable. Vaudeville seems to have lost out to motion pictures, perhaps perceived as better by audiences and cheaper by theater owners. While some media, such as printing and the telephone, exploded in terms of usage soon after their invention, others took a long time to develop. The facsimile and computer were both invented in the early 19th century, the former by Alexander Bain, in England in 1843, the latter by Englishman Charles Babbage in 1834. But there was no electrical communications network for fax, and the computer's inventor could not get his model produced in working form. So, some make it instantly, and some take a century to catch on. But we do have to recognize that they interact. Some depend on others, some merely compete with others. That is why it is so hard to predict.

It is interesting to note that the fax, the computer, and the telegraph were all conceived within a relatively short span of time in the early 19th century. Near the end of that century came the discovery and harnessing of electromagnetic waves, to be eventually used in radio and television. These were periods of great inventive progress, even if the practical use of the inventions may have taken some time to materialize. What was lacking in those times was the interconnection of the inventions — sending fax images by wireless, or television images into a computer. So, as Negroponte suggests, it seems to be the combination that is accelerating our own era's developments.

There has to be a limit to media growth, of course. We cannot keep increasing our use of media. The length of the day is still 24 hours. It may well be that leisure has stopped increasing for most of us, and this means that any new media growth has to come at the expense of older ones or of other forms of recreation and instruction. Regretfully, the individual consumer has little to say about what gets developed. We do, however, have a say in what products we buy, and that can be a powerful influence.

Chapter 4

The Special Place of Books and Writing in Our Culture

Book lovers often react negatively to the concept of electronic books or journals. One reason is typical overselling by techies. Another is our dedication to an image of the book that may not be all that realistic. The history of what we think of as our literature (defined in my dictionary as *writings*...) began with the oral tradition, works intended to be conveyed to an audience orally, not in writing. Gutenberg gave us the beginning of the modern book, but I wonder how many who cherish taking a good book to bed would like to take a Gutenberg book to bed. Hard on the rib cage, a book that size.

My point is that books, now meaning the physical objects, have been changing since the day they were invented. And there is no reason to think we have reached the end of this process of evolution. The book considered as a manifestation of the thought of a human being has not changed so much since the days of oral culture. But it has changed very much physically since then, even though very gradually. Now books are fixed in content. Oral compositions couldn't have been so fixed. To which version of the book are we devoted, book as object or book as thought?

Printed books are technological artifacts. This technology was invented by people. Books are not part of nature. Their design at any given time is based on the combination of needs and preferences (including those "needs" created by marketing), materials available, technology available, and economics.

Book technology has changed quite considerably since the first codices appeared in the second century and it will continue to change, with or without electronic media being involved. What we have improved since the beginning are such features as more readable type styles, lighter-weight page and cover materials, lower cost, inclusion of graphics beyond first letter illumination, spaces between words, division into chapters, and addition of tables of contents and indexes. New materials and their costs, the costs of production, social changes, and consequent changes in user demand will dictate future changes, just as they have in the past.

Although of course I can really only speculate, it seems likely that most major changes in the past were met with the same initial distaste by readers or hearers being expressed today. Jacob Burckhardt described this early attitude toward printed books:

> The material used to write on, when the work was ordered by great or wealthy people, was always parchment; the binding, both at the Vatican and at Urbino, was a uniform crimson velvet with silver clasps. Where there was so much care to show honour to the contents of a book by the beauty of its outward form it is intelligible that the sudden appearance of printed books was greeted at first with anything but favour. The envoys of Cardinal Bessarion, when they saw for the first time a printed book in the house of Constantine Lascaris, laughed at the discovery "made among the barbarians in some German city," and Federigo of Urbino "would have been ashamed to own a printed book."[1]

Not much different from the view in Victor Hugo's fiction, quoted in Chapter 1. Federigo, the Duke of Urbino, 1422-1482, was a collector of books, many of which he paid to have copied by hand. How different was his thinking, in an age when the handwritten book has virtually disappeared, from, "I would not own a book on disc?"

Not all books are alike, nor should they be. The publication medium currently best for a detective story or Harlequin romance may well be the pocket-sized, paperback book. Not so for a compendium of current tax laws, which have the distressing tendency to change often and with little notice. Not so either for an encyclopedia, expected to have a shelf life of decades.

Most, or perhaps only many, present-day people have a reverence for books, even if they are not readers themselves.

What, After All, Is a Book?

Aidan Chambers described an unexpected experience while teaching an undergraduate course on children's literature. He asked his class what a book was. "'Define a book for me,' [he] said, expecting an instant answer. Nothing." Mr. Chambers is an author of young peoples' fiction, a university lecturer, a publisher, and a devotee of the theater. In short, a lover of books and a person unusually well qualified to comment on them. I, on the other hand, while also a lover of books and a university lecturer, am merely an unschooled lover of theater and an author of nonfiction books. I am often accused of being a "techie," not without justification.

The answer to his question that Chambers and his class eventually worked out is:

A book is a sequence of pages on which appear meaning-communicating marks, all of which are bound together in an authorized order.[2]

The use of the word *authorized* here seems to be an intended pun — the order is specified by the author, hence is *author*ized. The word *author* comes from the Latin *auctor,* which, in turn, comes through a linguistic chain from a word meaning to augment or increase. *Auctor* implies an originator or promoter rather than a person whose words are canonical, but apparently the idea took hold that a person whose words made it into print was worth paying attention to. That idea still remains, but may die out, pushed along by desktop publishing, which tends to carry no promise about the reputation of the author.

I was surprised by this definition. Here is an eminent humanist taking a rather technological view of a book, while I, the technologist, regard the essence of a book as a *work*, more specifically a recording, whether textual, auditory, or graphic, of the work of a human being. (Yes, there is the story of the monkeys, set to typewriters, who could eventually produce all of Shakespeare, but nonhuman works are not a class under consideration here.) To me, the manner of recording is distinctly secondary to the content. While I expect technology to induce changes in the recording medium, and this to induce changes in the way works are created (à la McLuhan), I most certainly *do not* expect the act of creation to cease. Great works will continue to be produced and read or otherwise perceived. Most will be suited only to their age, some will be ageless. These latter will undergo many media transformations over their span of existence, as they always have. Perhaps we might consider that the definition of a classic is a work that retains its appeal even as the dominant media in a culture change.

We hardly have vocabulary with which to express the generality of creating and comprehending the works of the human mind that do not relate directly to writing and reading. Words such as *writing* and *literature*, while having multiple meanings, are all derived from the use of alphabets and books. There are really several ways to regard the book. It starts as a construct in the mind of an author, who may also be a graphic artist or musical composer. That construct is then recorded, traditionally, as print, musical symbols, or illustrations on paper. The book is then read and every reader makes, or is free to make, an individual interpretation of the symbols. With this view we can expand Figure 2 to show in Figure 3 how both originator and destination rely upon prior knowledge to create and interpret messages.

Barthes distinguishes among the roles of scriptor, reader, and critic.[3] I am inclined to view the critic as a reader who writes a new text, linked to the old, thus starting a new cycle by becoming a scriptor. I prefer to use the expression *work* for the mental construct of the author, which may then be actualized or represented as graphics, sound, or text. The work is then transformed into a *reader's* image or *mental model*. A work is created and becomes a book. Along the way or afterward, each editor, critic, or reader creates his or her own image of what that work is. (One of my favorite criticisms of critics is their tendency to review the work they seem to wish had been written, or that they themselves might have written, rather than the one that *was* written.) Further, the manifestation of the work may change as it is reprinted or excerpted. How much does the form affect our sense of the content? I believe quite a bit.

Not everyone would agree that multimedia books are books in the traditional sense. But they are works. It is my belief that more and more books will be produced in this form. It is also my belief, and hope, that the pure text form, whether recorded on paper or computer discs, will be with us for a very long time to come. However, they may not be with us forever.

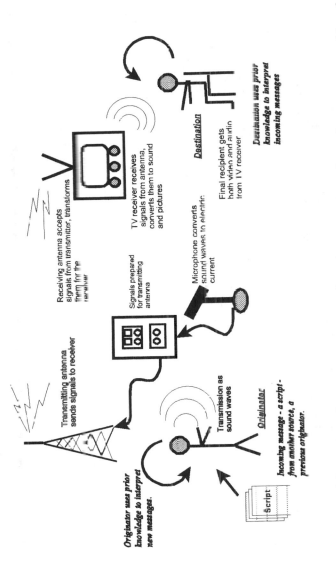

Figure 3. Shannon's model extended. *The person speaking and acting receives input from other sources, uses his or her own prior knowledge and experience to interpret them, and decides how to present them on camera. Similarly, the destination relies on prior knowledge to interpret the incoming messages.*

The Sequence and Discipline of Reading

Traditionally, authors produced text as words in a certain sequence and presumably expected their readers to read in the same sequence. The work may not have actually been written in that sequence; it may have been written piecemeal and reorganized often during the process of creation. Traditionally, readers take what liberties they wish with the order but in reading novels we generally follow the author's order, while with reference books we are not expected to. Hypertext is a mode of writing in which it is the author's intent that readers will select their own sequence or that the sequence will be determined by the reader's responses to questions posed by the author. Proponents argue that this enables the reader to become involved in deciding what the work is — it could be different for each person. Others worry that encouraging readers not to follow the author's line of thought can cause loss of intended meaning or effect. More about this in Chapter 6.

Some nonfiction works are likely to be so constructed as not only to permit but to encourage nonsequential reading. The *Handbook of Physics and Chemistry*, an invaluable collection of tables of mathematical functions and physical measurements, would be relatively useless if it had to be read from the beginning all the way to the item a reader wanted to find, such as the melting point of an element or the cosine of an angle. Encyclopedias and dictionaries are obviously intended as more or less random access devices (which doesn't mean that people use them randomly but that the time to access any location selected at random is relatively constant).

Along comes hypertext and with it the McLuhanist belief that linear, sequential reading is *not* the thing to do. In hypertext, text passages are presented in small segments of about one-half to one-quarter of a typical printed page. The reader is given a number of options on how to proceed. Jay Bolter gives a strong argument for the value of hypertext fiction.[4] It requires that the author consciously plan the work this way, but the result can be a book able to provide a different experience for almost every reader and even for the same reader on different occasions. In segment one, John meets Mary. The reader may decide to read more about John, or Mary, or what brought them together, or what happens next. The reader participates in organizing the story's presentation, even though the author wrote it. The reader's experience can be significantly different on a subsequent reading. Different, yes. Worse, or defective, compared with linear fiction? That answer lies with the reader.

Similar things have been done with the kind of multimedia "books" that are works played through a computer. There can be full motion video and sound, as well as "printing" displayed on the viewer's screen. The viewer can have whatever degree of choice over the sequence of presentation that the author chooses to give. I was recently fascinated to watch a young cousin of mine, age about four, completely enchanted as, providing inputs along the way, she watched a mult media story unfold. She could also have chosen to have the narration in English or Spanish.

Such works place significantly different demands on the author. They are different media than we are used to. They are not books in the usual sense, not audio plays, and not motion pictures. They are all three, combined with the computational capability to elicit input from a viewer and act upon it to change the sequence of events. We are going to see more of these productions and, eventually, they are going to be seen as an art form. But think how long it was after the inventions of radio, phonograph, and cinema before anyone considered them as art media.

Theodor Nelson and Douglas Englebart are generally acknowledged as the inventors of hypertext. Nelson suggests as the ultimate form a text wherein every word offers an opportunity for a variety of possible next words.[5] So, total freedom of sequence would be one possibility for book production, but with it would come the obligations to make frequent decisions and to use a sequence that would lead to the desired outcome. A desired outcome would, presumably, be finding a "fact" or learning what an author thought. Can the reader judge the likely consequences at each point? Does the reader want to be bothered?

This brings me to consider a very old-fashioned idea — discipline. Is the total freedom we can now offer readers, and encourage them to use, a universally good thing? Is it not better to at least encourage readers to follow the author's intended sequence? And for most works, is it not far easier for an author to follow a single sequence? Indeed, if we really want to understand, isn't it essential to resist page-hopping and read the book as presented? Again, if the author *wants* the reader to exercise full control, that is a different story (another pun intended).

Ivan Illich refers to the need for discipline in reading by which he meant the discipline of paying respect to the author — that we owe it to the author to follow the intended sequence in order to follow the author's line of thought.[6] Speaking of sequence, Illich mentions a 12th-century author who alphabetized portions of a reference text on biology and then apologized for using the alphabet, rather than an inherently meaningful sequencing basis, such as the type of animal or plant, as a way of ordering the presentation.[7] Here is a case of the author assuming readers want to have their information provided in a meaningful sequence but taking the easy way out and using a largely meaningless sequence.

One negative aspect of loosening readers' discipline is the possibility that future generations would become snippers of items, rather than comprehenders and deep thinkers about what they read. We'll go into this more in Chapter 11.

To sum up this argument, handing over sequence control to the reader, or doing so partially, is neither an unmitigated good nor evil. It is a tool available for those authors and readers who can, and choose to, use it.

Other Forms of Writing

While this book is principally about books and their modern variants, we must also recognize other forms of writing that may affect our attitudes about books. Letter writing was once an art form in itself, but seems to have declined significantly in popularity. At one time, in fact for quite a length of time, letters were the only feasible way to communicate between people widely separated in distance. Further, in the day of handwritten letters, the writer had to think carefully before committing words to paper. They were erasable only with great difficulty and too much scratching out looked bad. Today we are more likely either to use the telephone to follow AT&T's dictum, "reach out and touch someone," or to use electronic mail. Other than in connection with the Watergate and Irangate scandals in Washington, and Prince Charles's regrettable affair in England, it does not seem likely that *The Collected Telephone Tapes of X*, or *Y's Contributions to the Listservs* will be published as art. Or is it all that unlikely? In another generation, these may seem as natural as *The Collected Letters of William Butler Yeats* seem today. We must also remember that early writing was mainly for commercial and official purposes. It had to take time for people to learn to use this mode as a means of communicating their personal feelings.

The informal note is another form of communication many of us cherish. In what may be a rare reversal of a trend, the Post-It™ seems to have increased our usage of this mode of communication. Notes, like more formal texts, could not have been used much in the days when most people could not read or write. Today, the little yellow or pink slips are everywhere. Notes are used for short communications, probably not for deep expressions of feeling. The simplicity and convenience may be hard to replace, but voice mail and e-mail could compete. Voice mail messages can be recorded so as to say, "If it's Jane, please meet me at the library," and door bell systems could be programmed to deliver the same kind of message in response to a ringing code prearranged with certain frequent visitors, instead of the resident taping notes to the front door for all, not just one addressee, to see.

Culturally, the generation before e-mail and relatively cheap long-distance telephoning could put a great deal of feeling into a letter and perceive feelings as well. This is, or at least I believe it is, due to cultural conditioning. If people do not learn to express their feelings in writing, they cannot do so. If people are not used to the written word, as opposed to the video or cinematic word, to convey feelings to themselves, they might not be able to extract the emotional content from a text. Similarly, I wonder whether telephone users of, say 1890, ever put any emotion into the telephonic conversations. Modern people tend to laugh at the artificial emotionalism of 1930s movies. The directors seemed either not to know how to express emotion realistically or assumed their audiences preferred the artificial and stilted form. The viewers of that day did accept the directors' representation of life. How would a 1930s viewer react to today's cinematic depiction of love or war? Would the American Revolution and the subsequent creation of its Constitution have occurred without the stirring or convincing writings of Paine, Jefferson, and the Federalists?

Gatekeeping, Censorship, and Responsible Editing

Publishing a book or article is, in itself, an act of respect for the author. Not that garbage is never published, but it costs money and effort to put out a book, it involves some risk of both the publisher's money and reputation, and someone had to be convinced it could be worth a buyer's time to read it. Self-publishing, now made easy and inexpensive by word processors, makes the act of publishing appear deceptively easy. Among the consequences of desktop publishing and the uninhibited Internet is that it is becoming no longer a mark of respect just to be published.

One aspect of the brave new world of telecommunications is quality.

The term *gatekeeper* is often used in journalism and information studies to denote a person who controls the flow of information. It can be almost pejorative. But let us remember that the fictional, irascible, and beloved television and newspaper editor Lou Grant was a gatekeeper. The *New York Times* editorial staff is a collective one, as is Alfred A. Knopf, and the University of North Carolina Press. I respect these institutions very much. I rely on their good judgment. They must not only limit what they publish, but decide what the public wants to see and they must give some consideration to what the public *should* see. They are not infallible, which is why we are fortunate that they have many competitors. Competition does not mean only vying with others for the readers' money, but for their attention as well. Thus, government agencies, political parties, and nonprofit organizations compete for readers just as do commercial publishers. Another aspect is that the commercial competitors are buying each other up at a furious rate with the possible consequence of reduced variety of offerings to the public. I also worry that there is too much concentration by Internet aficionados on the mechanics of distribution rather than content. There is little, if any, quality control on Internet content.

So gatekeepers are people who decide what information shall be made available through the organization for which the gatekeeper works. They are known by such titles as city editors (in newspapers), acquisitions editors (in book publishing houses), acquisitions librarians (in libraries, of course), teachers, and parents. Sometimes they are known as censors. And, indeed, while part of the job of most gatekeepers is economic, another part has to do with quality and taste. When these last words are used, the word *censor* often follows quickly. When the gatekeeper rules on the basis of point of view, we generally call that censorship or protection of our children from foul foreign influences, depending on point of view. Even rulings based on quality can lead to charges of censorship. Rejected authors do not look with kindness on their rejectors.

Publishing books in electronic form does not change the situation regarding censorship or gatekeeping. It still costs money to publish a book, publishers must still make decisions about where to invest their money, and they still will value their reputations. But on the Internet, the economic situation is quite different. Once a manuscript is written (the author's investment) it costs relatively little to get it on the Internet, by comparison with paper publishing, and there is at this time very little in the way of controls on content. There have been charges of gross obscenity. In Canada, the details of some grisly murders and the subsequent, separate trials of the two accused were banned from the news media prior to the second of these trials.[8] The news, however, was fairly freely available on the Internet through on-line newspaper services originating outside the country and therefore not subject to the jurisdiction of Canadian courts.

On the other hand, the Internet has been called the most democratic institution we have in North America, precisely because of this complete freedom of access. We have, then, a conflict probably as old as society. Is it right, proper or fair that some people exercise control over what others may see, hear, or read? Another way to view this question has been raised by Daniel Boorstin. Is there a Gresham's Law of information? In economics, Gresham's Law says that bad money drives out good. If there is some debased currency in circulation, anyone holding it will spend it before spending reliable, higher-valued currency. Boorstin put it that "in our ironic twentieth-century version of Gresham's law, information tends to drive knowledge out of circulation. The oldest, the established, the cumulative, is displaced by the most recent, the most problematic. The latest information on anything and everything is collected, diffused, received, stored, and retrieved before anyone can discover whether the facts have meaning."[9] If an information medium carries a lot of low-valued (admittedly a subjective term) information, will those with high-valued information contribute it to that medium? Would a research scientist or serious art critic contribute research papers or essays to the supermarket tabloids? Would that reflect positively on their work? If the Internet acquires the reputation of containing a large proportion of low-quality information, will high-quality contributors continue to use it? Will readers who demand high quality continue to use it?

If any of the questions just raised gets the answer *no*, then what is the remedy? Should there be a board of control over the Internet? Consisting of whom? Should the whole thing be commercialized and the market decide what gets accepted and distributed? Might that put the whole Internet under control of one or a few media giants? What should be done? I hate to cop out and fall back on an old saw, but as the variously attributed saying goes, "Eternal vigilance is the price of liberty." In other words, there is no easy solution. People who care are going to have to watch out for and even fight for what they believe in. There is an international aspect to this issue, as well. We'll discuss it in Chapter 14.

Reverence for the Book

It is always a risk to speculate on the reasons why people hold certain emotions. I believe that people in general feel reverential about books. There are exceptions: at a literary cocktail party where everyone is an author or critic, or on a university campus where we all publish books that no one wants to read except unhappy students given no choice. But for the general public, whether educated or not, whether or not they are readers, they respect books and respect anyone who has published one. This, surely, is an old phenomenon and it is no coincidence that *bible* comes from the Greek word for *book* and that early books were highly likely to be bibles.

That we also tend to respect ownership of books is evident in the way we proudly display them in our homes and offices. Who can resist, on first entering another person's domain, scanning the bookshelves to see what sort of person dwells there?

There is also stature to be gained from books we have read. A few years ago I was touring Jerusalem with my brother. We paused for a while outside the Dome of the Rock, atop what was once Solomon's Temple. An elderly Palestinian man struck up a conversation. In hesitant English he told us he was a plumber and that he owned many books. He then challenged my brother to say how many times he had read his holy book, clearly meaning read in entirety, not just passages. On receiving the answer "Three," he swelled with pride, and triumphantly announced his own score, "Six." Nothing was said about belief in or adherence to the laws stated in the books, but it was clearly implied that the reading itself conveyed some sort of blessing.

The physical appearance or construction of books changes relatively little in any lifetime. So it is also not surprising that when, for the first time since the 12th century, we are faced with a major change in book technology all at once, we cling to our beloved past. And it continues to be unsurprising that we get an emotional, negative reaction from so many people to the suggestion that books to sit down with and read might appear in a completely new form. But new forms of the books we use to look something up in seem already to have achieved acceptance.

What is surprising, at least to me, are the few computers that have so far been designed primarily or solely for reading books. Their most common feature has been a small screen, far smaller than the typical 25-line by 80-character computer screen. Then, there is the legibility, or lack of

it, under widely varying ambient light conditions. Why con't major electronics manufacturers recognize what this does to book readers? But then,
why were the first VCRs designed the way they were, able to be "played"
only by graduates of a video-game apprenticeship?

Add to all this the undoubted good qualities of present-day book technologies. The size, weight, legibility, quality of graphics, price, and durability of books all argue strongly for their continuance in our culture,
so long as it is still essentially our culture. Price may be disputed by
some, dismayed at the rising cost. But this book costs about three or four
times the price of a motion picture admission in Toronto. Yet (the author
said modestly) it provides more hours of fascination and there is no need
to pay for popcorn, baby-sitters, or parking. The content of a book such
as this is likely to lose value in a few years, as the world changes, but the
object itself may last for centuries now that publishers are again using
acid-free paper. The major threats to this ideal object are the cost of
paper and the value of being able to search rapidy within a text and copy
portions easily. The cost of paper could become an irresistible force
moving books from present form onto discs, and is discussed more in
Chapter 13. We will discuss copying and copyright in Chapter 14.

Apple Computer Company's Newton computer, a handheld device
about the size of a small paperback book, has (to me) a nice feel. Those
who like the feel of a good book could come to like the feel of a good
book-reading computer. Franklin Electronic Publishers' computer-reader
for the Bible is brown (leather colored?) with gold lettering for the keys
(gold embossing?). We will, someday, see electronic devices that are about
the size and weight of a 15 by 23 cm (6 by 9 inches) book, weighing only
a bit more, and with a good-quality display screen. This might be something like the foldout display panels often found with today's video
camcorders, which *are* intended for outside use. Then we may see some
competition in the hearts and minds of book readers, as well as in their
pocketbooks.

The Book as the Record

Some phrases common in our language attest to the value of writing as a
fairly permanent medium in which to record truth, law, or revelation.
These are some pretty heavy words. When we talk, we are apt to say,
"It's in the book," "You could look it up," or "It is written." One of the
great benefits of writing has been that words, or symbols, once written,

stay written, as have the *Dead Sea Scrolls*. The law is the law. Your school grades remain fixed years after graduation. Taxes are based upon the written record of earnings and expenses. Yes, of course, it's not quite that simple. Of course, as soon as words are written people start to debate what they really mean. But they will tend to return to the *text*, each side of a debate using the same record to prove its point.

Clearly, it cannot be this way in an oral culture. As the law and lore are passed down they are going to change somewhat. The dependence has to be on the *person*, the interpreter of law or custom rather than the immutable law itself. Although different from our way, there is nothing particularly wrong with this. You have to trust the interpreter but in our literate world you tend to trust the minister, priest, rabbi, imam, teacher, or judge who insists on telling you what the text really means.

Technology is now looming before us that can change the sanctity of the recorded word or image. For example, a picture, stored in a computer, can easily be modified without degradation of image quality. A text, too, can be modified as hordes of word processor users know. An article may be published in a scientific journal or newspaper and a reader may include some of its content in a new article of his or her own (giving due credit, of course). Then, suppose the author of the original finds an error and changes the computer-stored version of it. Future readers might see only the new version unless someone makes the effort to keep both old and new and to inform all readers that both versions exist. Even so, the authors who made use of the first version may now have errors in their work and may not know of the correction to their source material. We could lose the very concept of *original* or *canonical* — that is, of identifying a version of a work that is the first one published or accepted by an editor. A document, whether text, graphic, or sound recording, could become only what it is at the moment, not the same for all time.

Yes, there are ways to overcome such problems, especially by having some agencies, whether libraries, publishers, accounting offices (Which tax return is the official one?) be required to maintain each original in their domain, in effect under seal, unchangeable. Subsequent revisions, while still permissible, could be required to refer to the original and identify which change this is, say by an issue number. Using some modern facilities of the Internet, it would be possible that when a person read an article citing another, there would be an automatic search triggered to see if any cited articles had been revised and, if so, to bring them to the searcher's attention. But that does not work unless everything is on the net. And if *everything* were on the net, that could be very frustrating for searchers.

This is another of those areas in which the change from print to electronic recording *could* change the culture. Whether it will or not depends on whether we take appropriate steps to retain the original and to inform all users of subsequent changes that the changes exist. Or, we could not bother and switch to a culture in which the written word is not so sacred. That would be hard to do within one generation.

The main point of this chapter has been that the book occupies a very special place in our culture. Yet, we have quite different ideas about what a book is. Many readers feel that present-day book technology is ideally suited to our needs. There is a tendency to forget how much this technology has changed in its 1,800-year history. It is not static. It was probably considered ideal by some in 1455. It will change still more.

But whatever we mean by *book*, we tend to revere it. This undoubtedly grew out of the fact that early books were of a philosophical or religious nature and thus deemed worth revering. As their cost has come down and the market increased, we can hardly claim that all present-day books are worthy of reverence, but still they get it. Books — probably referring to content — are something special. Therefore books — their paper and ink embodiment — are revered and clung to.

As the technology continues to evolve, new generations will see the new forms of books as ideally suited to *their* needs, and the cycle will continue.

Chapter 5

Representing and Presenting Information

A common aspect of all media of communication among humans is their dependence on a combination of both signs or symbols and form or structure. Whether this be written language, a fugue, or a painting, effective communication relies upon basic symbols and the manner in which symbols are combined to form larger symbolic units. The basic units can be letters or words, musical notes, shapes, or colors. The larger units can be sentences, sonatas, or portraits. We get little communication if we must use only the basic units, with no means of linking them together to form meanings more complex than a single symbol can convey. The forms available to the creator, the author, composer, painter, photographer, or choreographer are to a large extent influenced by the media that will convey the images to a human.

Using language, whether oral or written, a writer can describe a scene. A really accomplished writer can induce a very vivid picture in the mind of the reader. Or a very imaginative reader can create the vivid picture from even a dull written description. But the author cannot literally paint the scene or capture the sounds of the forest or of the sea with words. The author has to use the symbols and combinations of symbols appropriate to the medium to bring out the reader's own perceptions and previous experiences and thereby create the desired effect.

As I pointed out in Chapter 4, to some extent the messages we can send and receive depend on our training and conditioning. A course on music or art appreciation can change the ways we perceive works in these media and, of course, most artists, composers, and performers have had extensive training in how to use their media. We can still appreciate music without the training, but are likely to miss much of the subtlety.

Until we began to invent modern information machinery, there were only quite limited ways by which to record information or works. The oldest of these, sound, had to embed the work in a human memory to record or save it, a method that we still use. There have been and still are purely oral societies. And I have to wonder, only half facetiously, whether pocket computer "reminders," cellular telephone, Post-Its, voice mail, and electronic mail might not drive the human memory out of existence from disuse.

Early on, earlier than we have a clear record of, humans devised tools for making retainable marks on a flat surface — a piece of red iron ore to write on a wall, a stylus for use with soft (later to be hardened) clay, eventually a brush or pen. With the recordings these tools enabled us to make, memory became partially mechanical. And thus, "truth" became relatively permanent, the same truth for any reader, at any time. This last statement, I hope you will realize, is facetious. I do not literally believe it, but a great many people do, although they do not necessarily agree on which recorded works contain "the" truth.

Over the years, the recording media became specialized: writing became based on the alphabet or a set of ideographs; graphic art became more skillfully representational, and then went beyond representational art to representations of the artist's impressions, not just the scene before the easel. The inscribing tool in the form of a chisel enabled carved sculpture. So, at about the time of Gutenberg, the recording forms were well developed but quite different from one another. We could record text, commercial accounts, two-dimensional art, music and three-dimensional art, each in its respective medium.

The human memory is the universal recording medium, universal in the sense that it is used to some extent in all communication among human beings. Machines and animals can send or receive messages in ways we cannot, such as by use of infrared or ultraviolet light, or high-frequency sound above about 20 kilohertz. An important question for the future is whether media are going to continue to diversify. If they do, and because they may require expensive machinery to enable humans to use them, the diversity could become self-defeating. Is there some way to reverse this trend? Do we want to? Forward media thinkers of the day are suggesting that we can reverse the negative aspects of the trend without giving up the diversity. Before we consider how, let us explore how we represent information in the major media of today.

What Is Meant by Information Representation?

We have to start with the reminder that the printed word or the painted picture is not, itself, information. The information is the image that is created in the mind of the beholder or the effect created upon receipt of the message. It never ceases to amaze me how diverse and vivid are the images that can be created in our minds by relatively simple tools — pen, vocabulary of as few as 3,000 words coupled with syntax, or, in the pictorial world, a brush, paints, and the "syntax" of painting. Syntax of painting? There are rules about how paintings are composed, how colors are mixed, how colors complement one another, how perspective is represented, where the principal images are to be placed within the boundaries of the picture. Both artists and viewers have to learn something of these rules or conventions. Would an art lover of ancient Egypt, where the graphic images always seemed strictly two-dimensional, have understood what was being conveyed by one of our modern graphics with the third dimension so "clearly" portrayed?

I have had a number of cats in my lifetime. Only one ever showed any interest in the television, except as a warm place to sleep. That one used to sit on top, look down onto the screen, and try to catch football players. There seemed to be something about the relative size and speed of football player images that struck her as interesting and worth pursuing. I have never seen a cat show any interest in a TV image of another cat or in its own mirror image. Cats seem not to have learned to see these images as creatures like themselves. Or perhaps they are multimedia creatures and must have the smell as well as the picture to recognize another animal.

Representing Text

We represent words by a sequence of letters and these, in turn, represent sounds. Whether originally each letter represented one and only one sound, we cannot know. In the languages I know, letters may have more than one sound and the same sound may come from more than one letter. Eventually, we added ron-acoustic symbols to the language such as the space, capital and small letters (upper- and lowercase, in the words of cold-type printers), and punctuation. Some punctuation can be used to signal inflection or tone, as in the difference between "That is a dog." and "That is a dog?"

In writing, with handwritten letters, the letter shapes are copied from memory onto a page at the moment of need. In early forms of printing, which lasted until this century, we used the same letter shapes but their appearance at a given position on a page was assured by inserting a piece of metal type into an image of the page. The piece of type was, in effect, a pre-written letter. All the letters for a page were assembled in the correct order and then ink was applied to them all, essentially at once. Then, a paper was pressed against the inked type and the whole page printed, essentially at once. I say "essentially" here because there were, and are, many variations in how type is inked and the paper fed to the press.

With the typewriter, as in printing, the letter is preformed. But, as in writing, the letter forms are applied to the page and simultaneously inked, at the time and in the place where the letter is needed, one letter at a time.

When the computer came along, the manner of recording the text inside the machine was and is quite different. The representation of a letter inside a computer bears no resemblance to the letter as it appears on the printed page or to the sound it makes when spoken. In this sense, the computer is like the human brain, in which we have no reason to think that whatever it is that tells us what A is must look like A. The letter A in most computers is represented as 0100001 where the 0s and 1s are binary digits, or *bits*. They are, in turn, represented not as digits but by the electrical state of tiny circuit elements that cannot be seen or felt by humans.

When the old-fashioned human printer wanted an A to be printed, he or she would reach into a type case (upper for A lower for a), remove a piece of type, and set it into the image of the page being built up. When a typist wants an A, he or she strikes the A key (for a) or the combination of *A* and SHIFT (for *A*). When a computer of just a few years ago wanted an *A* to be printed, it had to convert 0100001 into instructions to its printer (now meaning a machine, not a person) where to apply a bit of ink to a page. Early computer printers were like typewriters, which used impact printing and typefaces built into the printer. These are now rare.

The modern computer printer reverts to the age when the letter did not exist, except as a generic mental model, until the moment of writing it on the page. The modern machines do not draw the letters as we would (as I was taught to, years ago, in an engineering drawing class): To make an *A*, make the mark /, then \, then –. Today, it is more like the sequence shown in Figure 4. This illustration is exaggerated. It actually requires more "scan lines" per printed line to give good-quality, what is now called "letter-quality," results. And the scan is all across the page, making the top of every letter, then the next line for every letter, etc.

If the computer tells its printer to print "Hello," the latter will easily do it. The computer (instructed by its user) can also tell its printer to print in italic (*Hello*), boldface (**Hello**), or script (*Hello*). These variations come from a generic representation of "Hello" in the computer, plus extra in- struction on what type font to use during the printing process. So, the *presentation* can vary considerably once the basic method of *representa- tion* is established.

Another way to bring the text out of the computer for presentation to a human is to convert it to sound. Today, many of the messages you hear from your telephone company ("The number you have dialed, 555-2345, is out of service. The new number is 555-5432.") are not spoken by a person, but by a computer voice synthesizer. The same synthesizer can convert 0100001, one representation of *A*, into the sound "ay," another representation. If well done, the synthesizing will consider context and so not pronounce the *T* in *acting* the same as the one in *action*. Some synthesizers can also do regional accents. This method, then, gives us another variation of presentation, from the same basic representation. In general, we store a code in the computer representing a letter. We must also store information about how that letter is to be presented, as part of a printed text or of spoken words. If printed, in what type style and size? If spoken, with what loudness, pronunciation, and inflection? Another way to look at this is that representation in memory, whether human or machine, may and in general will be different from the representation needed to convey the image to another person or machine. We are calling

Figure 4. Forming letters with a modern printer. *This is a schematic of how printing is done by an ink jet or laser printer or a fax. Here, we see the first four lines, then the complete letters. In one scan the very top of the high letters is put down, and subsequent scans fill in the rest. In actual practice, more scans may be used per line of print than are shown here, resulting in a clear, sharp image of the letters.*

that transfer from one storage to another *presentation.* Computer people may call it *display* but that seems not general enough when we want to include sound and even touch as options. Presentation, then, means transforming a representation to assure or assist intelligibility by the recipient.

It is this capability to transform representations that leads Negroponte, Eisenhart, and others to suggest the all-digital world in which there might be one standard way to represent information in computer storage and any number of ways of bringing it out and presenting it to the human recipient, the method to be used being determined by the human at the time of need.[1]

A Diversion: Analog and Digital, Characters and Pixels

Just to dig a bit deeper into how information is represented in electronic machines, we need to review four fundamental concepts. The word *analog* in computing or information representation means that one measurement is used in place of, or as an analog of, another. Probably the most commonly known example is the clock, the most traditional form of which, for most of us, is now called an analog clock. It uses the angle by which a mechanical arm deviates from the vertical to represent elapsed time. If the larger hand has moved one-twelfth of the circle, or thirty degrees, from the vertical, then we know that one-twelfth of an hour, or five minutes, has elapsed since the hand was vertical. A watch using *digital* representation, on the other hand, displays numbers, not angles. It may show 4:05, directly telling that five minutes have elapsed since four o'clock.

Most automobile speedometers are analog instruments. The angle of the pointer is analogous to the speed of the vehicle. The car's odometer, which tells distance traveled, is digital. It directly tells the number of miles or kilometers since it was last reset. In telephone transmission, the loudness of the input sound causes the voltage of the electrical output signal to vary analogously. That voltage variation at the receiving end causes a membrane in the receiver to vibrate and thereby to re-create the original sound.

When a digital clock says 4:05, its next signal will be 4:06. It cannot tell us how many seconds have elapsed since the 4:05 was first displayed. It is possible, of course, to have a display for seconds, so the time might read 4:05:13 when 13 seconds have elapsed after 4:05. Digital representations can be highly precise but this must be planned for in advance. An analog clock always allows the viewer to estimate how much after 4:05 it is.

Another example of some of the difficulties of digital representation is the infamous Year 2000 problem. Because decisions were made years ago by many software designers to use only two digits to represent a year, the year represented by 00 is now ambiguous. It could mean 1900 or 2000. In the year 2000, there will be both people whose birth years are 1900 and 2000. It seems trivial. It could be very expensive to fix.

There is yet another way to represent information for computer storage. It is similar to the way photographs are printed in a newspaper. When a photographic print comes back from the drugstore's developing and printing service, you can use a magnifying glass to enlarge your view of an area and see more detail. The more powerful the glass, the more detail you can see, until you go to the extreme of using an electron microscope to see the individual molecules of the photographic emulsion. Any further enlargement shows you more about the molecule, but not more about the subject of your picture. You have reached the limit. In a newspaper photo, if you enlarge it, even with an ordinary low-power magnifying glass, all you see is larger dots. You cannot increase the amount of information available to you by enlarging.

Computer people call these little dots *pixels* or picture elements. You can control the detail or resolution of a picture in two ways: by the size of the pixel and by the number of levels of gray tone or color each pixel can represent. Using more, smaller pixels gives a finer resolution and can represent more detailed information. If a pixel could only be pure black or pure white, a reproduction of a photograph would seem harsh. If each pixel could represent eight levels of gray, from pure white to pure black, the resulting print would be softer, capable of representing more of the true tonal variation of the original scene. If pixels can represent color as well, we get even more of the original in our reproduction. Today, we can go to thousands of gray levels or millions of color variations, giving computer output that is as fine as the best art book reproductions. Your $200 home printer may not give such results, but it can be done.

There are, then, two ways to store a document in a computer. In one, the characters are represented by digital codes, like 0100001 for A. Using this method, the computer can search for specific words, even analyze the subject matter of the text. The second method is to store a picture of the document, using pixels. The pixels are represented by digital numbers, but not the individual letters of the text. With this method, the computer cannot analyze content.

When text is represented in digital form, anyone interested can determine that the text contains certain letters, words, or phrases. By inference, if it contains these phrases, it must be about this certain subject. This section of this chapter uses such words as *analog, digital, pixel, representation, image, picture,* and *text.* A reader who knows some computer science could tell from these clues alone roughly what this section is about. But when a scene or face is represented by a set of pixels, we do not yet have the technology to scan the arrays of pixels and tell that, yes, this is a picture of a flower, or of a woman smiling, or of the sea in a storm. We may know there are so many black pixels, so many white, and so many of each gray level in between. But what do they *mean?* We don't know. People are working on it, but it is an enormously difficult job to determine what is in a picture by examining pixels, unless we severely limit the kinds of pictures we will deal with. We can, for example, recognize that the image of a fingerprint taken at a crime scene matches one stored in the police files. But a fingerprint is nowhere near as complex an image as is, say, a Marc Chagall painting. Progress has been made, roughly, in allowing a person to query a pictorial file asking if certain characteristics are to be found therein. That is different from asking what is contained therein.[2]

Representing Sound and Pictures

An early telegraph worked basically by sending a continuous, non-varying electric current from one point to another. It closed a circuit so the electricity could flow, then it opened (broke) the circuit to stop flow. The sender used a switch or key to break the continuity in a patterned way. The key made a noise as it operated, confirming to the operator what had been sent. At the receiving end, the opening and closing of the electric circuit activated and deactivated an electromagnet, which moved one piece of metal against another, also making a sound. Alternatives were to make a mark with a pen or turn on or off a light. The patterns were determined by Morse code. An operator had not only to know the code, but to be proficient in using it, i.e., do the timing just right so a group of sounds was unambiguous at the receiving end. The basic concept is digital — what is sent is a dot, a dash, a space between letters, or a space between words.

Alexander Graham Bell and others dreamed of sending voices over such systems, but this required that the transmission not be a continuous tone, but a varying one. By analogy, we could use our vocal chords to produce only a single tone. Repeating it different numbers of times or holding it for different durations between sounds could make for intelligible signals. But by varying the tones we can produce a far richer form of voice communication. Even birds, dogs, and cats, with limited vocal ranges, vary the tone of their utterances to produce different meanings. As we know, Bell was successful in sending an analog of the voice or other sound over telegraph-like wires to a receiver. This, of course, is analog transmission.

Radio was initially conceived as wireless telegraphy, but quickly also began to be used for varying sound transmission, now its virtually exclusive preserve. Radio works on the same principle as the telephone. The transmitter converts the input sound to electromagnetic waves, varying in frequency (frequency modulation — FM) or in amplitude (amplitude modulation — AM) and the receiver converts the signal back again into sound, using the same kind of speaker as the telephone. with a vibrating diaphragm. The electromagnetic waves travel, without benefit of wires, over great distances.

The original phonograph worked much like radio. Variable grooves in the record caused a steel needle to vibrate. This in turn created a varying electric current that was transmitted, within the machine, to an amplifier and speaker. So, conceptually, the only difference was that the electric current analog of the input sound remained a current, rather than being further converted into electromagnetic waves. Until very recently, sound recordings for phonographs have been analog in form.

Radio and phonograph are both gradually going digital. This means the sound is represented digitally by the equivalent of pixels. Instead of the sound element being coded for the color or gray tone it represents, digital radio gives the frequency or set of frequencies of the sound being sent or played at the moment of sampling. One major advantage of this method is that some degree of error detection and correction can be built into the system, so that external noise does not necessarily distort the intended sound, as it would in analog recording. This is especially valuable when copying recordings. There need be no loss of signal or introduction of noise when copying a digital sound recording.

Television is bundled with facsimile and photocopiers in its basic techniques. There is a camera or equivalent that scans the image to be transmitted. This might be a page of text, a still picture, or a wide-angle view of a football game. The scan goes from left to right and top to bottom of the image. At points along the scan, the input device detects the level of gray tone or the color of what it "sees" at that point. This is similar to recording pixels. After the scan, the scanning device moves down to start the next line. In the case of television, the scene is fully scanned 30 times per second. TV today actually transmits every other line of the image in $1/60$ of a second, the complete image then taking twice that long, or $1/30$ of a second. Faxes and photocopiers need scan a page only once, but they take longer to do so.

The TV color or gray values are converted to numeric values and an analog of that value is transmitted to the receiver. The fax does the same. Some faxes today digitize the data into discrete pixels before transmission. It would be possible to take the representation of a television image of a single frame, convert it to either the fax or photocopier form, and then print it as a still picture. The information needed for video, fax, or photocopying is not all that different, but neither is it identical. It could be unified, leaving it to the receiver to determine how it would be presented to a user. In television, the sound is separately recorded and transmitted as an FM radio signal.

Video or still photography can be used to record the image of a text, a fact useful in preserving paper publications. Newsprint is notoriously nonresistant to the ravages of time. Without microfilm, we would lose our old newspapers. But microfilm is not the users' favorite medium. So, how to preserve books that are beginning to disintegrate? One way is to digitize them and store the text in character form, from which they can be presented in nice, clear paper copy or retained digitally, or both. Most major daily newspapers do this, and this information is made available as a database, to be searched retroactively by those willing to pay for the service on the World Wide Web.

But it is expensive to digitize newspapers after printing. There are hardware-software combinations that can scan a paper, store the image in the computer as a set of pixels, then interpret the images as characters. This is called *optical character recognition* (OCR). But the character recognition systems now available are not yet good enough for the job of economically converting old printed material.[3] Stains and cracks on the page, poor type quality, and embedded graphic images all conspire to make automatic reading of the text difficult. Today we can put a photographic or video image of a page on a computer disc and help the reader with such software aids as turning a page at the touch of a key. But the computer cannot tell what the text of a video image of text is about. It is the same problem as trying to tell what is in a picture. So it cannot search for a particular word or extract a paragraph to be inserted into your own term paper or essay.

The manner of recording remains a dilemma for libraries intent on preserving their collections. Because of the typically poor quality of the print and the inclusions of graphic images, OCR is not economically feasible for large numbers of old books and periodicals. Scanning of text and its subsequent storage in graphic, not character, form is being used as a means of preservation of limited numbers of books that are in danger of decomposing or to help other libraries that may have lost a copy. The scanned images are recorded in pixel form, and then printed much as we do faxes. Once any book is scanned, it can be retained as a graphic image and reproduced for other libraries inexpensively. If the remaining original is valuable in itself, there is extra cost in scanning, because the pages cannot be disassembled and fed as a batch into the scanner. If many libraries, facing loss through fire or flood or future loss through decomposition, cooperate in creating an archive of scanned books, the cost of replacement for all can be lowered. At present, it can cost as much as 50 cents per page to reproduce a book, or as little as 10 cents per page if the book has previously been scanned. The latter cost compares with that of new books. Trying to buy a replacement for a hard-to-find or rare book can easily cost more than the copying price. Copying prices will surely come down as technology improves.

Bringing Them All Together

In the drive toward an all-digital, fully compatible world, we are making some progress but there is yet much to be done. A computer could receive a digital TV transmission, select the information that constitutes one frame, and modify it for fax transmission, photocopying, or printing on a computer printer. It is already possible to buy a single machine that serves as both a fax and a copier or these two and also as a document scanner and a laser printer. When I first priced one of these devices, the four-way machine cost about the same as the sum of four individual ones of each type. But six months later the price came down sharply and I bought one. This form of multi-presentation digital machine will very likely become a standard except for heavy-duty users.

As I have already pointed out, radio and television can be digitized and sent to a computer for presentation through its multimedia equipment. We can also go from digitized text to audio output, and have been able to do that for quite a few years. We can convert a page of text into video form for storage and later display. What we cannot yet do is any of the following:

Have a computer "look" at a picture and tell what objects or concepts are represented. We can do the equivalent with text, i.e., use the words and phrases to define the subject matter. But with graphics we can do this only in highly limited circumstances.

Scan a text, say of an old book or newspaper, and convert what is there to digital text. We can do this with good-quality text, but in the case of old, dirty, stained papers and poor-quality type, the results are not too satisfactory. One consequence is that it is still difficult to digitize old printing for preservation purposes.

We certainly cannot examine a text that describes a scene and then create a visual image of that scene. Nor can we create sound from a text description of sound.

One of the problems in moving from where we are today to the brave new all-digital, all-compatible world will be to resolve who does what. Which electronic devices will have to change? Which will bear the major cost of conversion and which will simply be converted by the others? No current broadcaster or print publisher is going to readily agree to make a major change in how their product is produced and delivered without a fairly strong guarantee of a healthy return on investment.

It is not likely that change will come all at once. On the other hand, incremental change, which always sounds so civilized because it gives us time to adapt, may cost more as we have to replace generation after generation of hardware as the recording and transmission forms change. But it does keep the economy hopping.

Chapter 6

Linear Text and Hypertext

Linear text is the only kind most of us know. The ancient languages — Hebrew, Greek, Latin, Sanskrit, and the modern ones derived from them — are considered linear in that we read along a line. Never mind if we go from left to right or right to left in rows, or top to bottom or bottom to top in columns. There is only one way to read the text in any one language. Usually, we do not know what is coming next, unless the writing is really unimaginative, nor can we jump ahead to where the hero gets rescued. We can jump directly ahead so many pages, but we have to sample the text to try to find the right content, telling exactly where the rescue took place, or exactly where a certain phrase is used.

My own view is that linear text has got something of a bum rap. Yes, the text flows linearly in a strictly technical sense. But the syntax has us relating words backwards and forwards. A pronoun can pop up quite far from its antecedent and, in spite of what our English teachers told us, in the days when English teachers told students anything about grammar, a pronoun does not always refer to the most recently preceding noun or pronoun agreeing in number and gender. Even when it does not agree, it can be and usually is still understood. In the preceding sentence, the word "it" occurs twice and refers to "pronoun" in the phrase "a pronoun does not ...." But, the closest preceding noun is "gender." We become very adept at working out these relationships at an early age.[1]

In a modern book, we may be given any of several kinds of tools to help jump around the book, or into it at a point of interest. The table of contents tells at what location broadly defined subjects are covered or, in a Dickensian novel, roughly what is going to happen to whom in each chapter. Nonfiction books usually have an index that tells where specific concepts or words appear. Large works may have a thumb index, as do many dictionaries, to give quick access to entries beginning with a given letter. Finally, the author may use footnotes, parenthetical expressions, or comments in the text to tell the reader where some other subject is discussed or to carry on a separate discussion that is tangential to the main topic under discussion.

The letters in English and other European languages go left to right within a word, the words go left to right on a line, the lines go top to bottom on a page, and pages are read from left to right, turning over the page when the bottom of the right-hand one is reached. Hebrew and Arabic reverse the letter, word, and page directions, but the effect is the same. By the way, in these languages numbers are read left to right. The year we designate 1998 in English is also written that way in Hebrew and Arabic. A good author, in using the language, does not have to be rigidly linear in terms of the sequence of what is described. The narrative does not have to be linear in terms of chronology. The order of describing persons, places, or events can vary. A Joyce Carol Oates novel, clearly a fictionalized account of the Chappaquiddick incident involving U.S. Senator Edward Kennedy, goes back and forth over the same material, each time going back a bit farther in time and forward a bit farther and also deeper into what is happening.[2] This is hardly a traditional linear novel. Yet, it is written in traditional text form.

Often a writer must implicitly ask the reader to follow a long chain of reasoning or narrative. Mathematics is like this, as are writings in most sciences. Mathematicians offer a host of definitions and proofs of theorems, all leading to a conclusion not always apparent at the start. The reader needs a certain amount of faith and the discipline that Illich speaks of. Similarly in a novel. Novels do not all start with the bloody corpse on the floor and the police sirens wailing. Sometimes it is very tedious getting started. Victorian novelists, paid by the word, were in no rush to get to the meat of things.

So, calling text written in conventional fashion in our language "linear" may be a misnomer. It is certainly an oversimplification. The basic symbols occur in a linear sequence, but the thoughts do not, or need not, occur that way.

What are the alternatives to linear text? *Hypertext* is a form in which a reader can jump around within the text on the basis of content, not just page number. *Multimedia* (the subject of Chapter 8) is a form in which information is passed from author's work to reader through several different media, usually more than one at a time. In both cases, both reader and author must adapt to a different form of information transfer.

What Is Hypertext?

With the traditional linear text, you read a paragraph or page, then go on to the next one, *next* in English meaning the one below, to the right, or on the other side of the paper. And there are the more modern ways of helping a person find an appropriate point, such as an index. In hypertext, after every segment of text, where a segment typically consists of what can be shown on one computer screen, the reader is offered explicit options such as to go to the next chapter, start over, follow the narrative, or find more about a character, place, or event. Or links may be embedded in the text by displaying certain words in a different color, typically blue. If the reader points to a blue word with the mouse, a program goes to certain text that contains that word. Sometimes these choices are annoying, sometimes delightfully helpful.

In nonfiction hypertext presented through a computer, for example a typical encyclopedia entry, the highlighting method is often used. These words indicate that they or information about them occur elsewhere in the document. Clicking on such a word will bring up the referenced article. If I mention Ivan Illich in my text, a reader might want to know more about him, right now. A click or two with the mouse and there would be the Illich article or text about him. Or a reader may find a word in a text unfamiliar. Another click can bring up a dictionary definition. These connections are called *links*. The text segments connected by links are called *nodes* or *pages*. The whole constitutes a large network of interconnected nodes. In a well-written hypertext, the author anticipates how readers might want to explore alternative paths and sets up the links. While links usually point to other segments of the same text, they may also point to other documents, which, in fact, is one of the primary features of documents on the World Wide Web (see Chapter 9).

I used the expression "well-written hypertext," and previously "a good author," referring to a linear writer. No matter what the technology, there will always be a need for creativity and skill in creating works that are to be read, viewed, listened to, or felt by readers, viewers, or, in general, *perceivers* of a recorded work.

Somewhat surprising to many observers, hypertext is not particularly new. Its origins go back at least 50 years. Vannevar Bush, who had been the science adviser to U.S. President Roosevelt during World War II, was concerned with the accumulation of scientific information as far back as the 1930s. As World War II came near its conclusion, the accumulation was enormous and continuing to grow. Bush proposed (but did not go so far as to design in detail) a machine he called *memex* in which could be stored a library of scientific information that could then be searched in a hypertext manner.[3] One concept might suggest another and so on, and these trails could be followed at will. Further, the scientist using the system could add his or her own thoughts, thus expanding the literature. Later, Douglas Englebart and Theodor Nelson worked on this idea in the mid-1960s and even gave the technique its name at that time.[4] Englebart acknowledged the influence of Bush's work on his own and Nelson's.[5] But Englebart's and Nelson's work did not attract much attention at first. Around that same time, computer-assisted instruction (CAI) was becoming the current rage in educational technology circles.[6] It was the technology that was going to revolutionize education. CAI used many of the same concepts as we now associate with hypertext, but in an instructional context. A unit of instruction would be presented, followed by a question. The next bit of text the student saw would depend on his or her answer to the question. If the answer was wrong, the next text could review the material or point out exactly what the error was. If the student seemed to be getting everything right, an accelerated sequence could be used. Before CAI went on a computer, the concept was embodied in printed books, called *programmed texts*.[7] Links were instructions to go to a certain page. One problem with this kind of book is that it did not serve well for review or for reference. But this flexibility is traditionally something we expect from textbooks. Its lack detracts from the value of this mode of instruction. Another problem was, and is, that there were not many skilled authors writing in this mode.

The Englebart and Nelson ideas became popular with the invention of personal computers and there is no doubt that its popularity is increasing.

Some people feel hypertext is the way of the future. Some hate it. Some see it as calling for more reader involvement with a text. Some see it as destroying creativity and comprehension, substituting instead skills in searching and copying. Some remember the hoopla over CAI, which has had a negligible overall effect on education to date, and wonder if hypertext will fare any better. Let us look, I hope impartially, at both sides of the issue.

For readers unfamiliar with hypertext, we have a conundrum. We cannot really demonstrate good hypertext in conventional print. These readers might explore the illustration in Figure 5, which is really a schematic, not an actual hypertext. More information about hypertext is cited in the notes.[8]

Advantages of Hypertext

Consider nonfiction first, as this is the domain in which hypertext has had its greatest success. A number of encyclopedias, traditionally published in print form, have brought out an edition stored on a compact disc, and using hypertext concepts. Included are such general publications as *Encyclopaedia Britannica*, *Compton's Encyclopedia*, Microsoft's *Encarta* (based on *Funk and Wagnall's Encyclopedia*), the *Canadian Encyclopedia*, *World Book*, and *Academic American*.[9]

There are also a number of specialized publications, such as Microsoft's "edutainment" productions, *Multimedia Beethoven* and *Art Gallery*, which are specialized encyclopedic tours of a subject. They can be followed in sequence or the user can browse at will. Among dictionaries are Merriam-Webster's *Collegiate Dictionary*, the *Oxford English Dictionary*, and the *American Heritage Dictionary*. What all these publications have in common is that they are intended to be used as reference works in which a user typically looks up a single concept and hopes to retrieve a short article. *Encyclopædia Britannica* (EB) has many lengthy pieces but the other works tend to deal with articles averaging closer to one or two printed pages or even less. EB breaks up their long articles into a series of shorter ones to enhance reading on the screen and finding the specific information one is searching for.

Often, a user does not know exactly what is wanted or exactly how to express the concept he or she may want. Or, especially with new technological terminology, the terms may not even be in the encyclopedia or dictionary. Where will I find material on hypertext? Under what title or

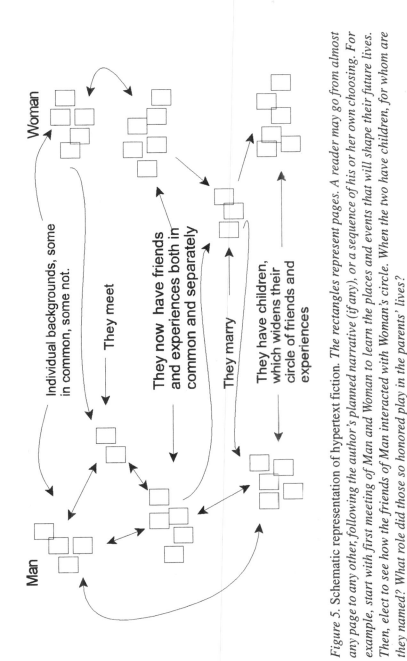

Figure 5. Schematic representation of hypertext fiction. *The rectangles represent pages. A reader may go from almost any page to any other, following the author's planned narrative (if any), or a sequence of his or her own choosing. For example, start with first meeting of Man and Woman to learn the places and events that will shape their future lives. Then, elect to see how the friends of Man interacted with Woman's circle. When the two have children, for whom are they named? What role did those so honored play in the parents' lives?*

subject? If I cannot find it directly, do I look under *publishing* or *computers*? (In an actual test, I found the term well defined in one encyclopedia. In another, it was not present and there was nothing offered to help me.) A search for information may involve hopping around, which can be bothersome in a printed multivolume work. With hypertext you can enter the word you want to start with, retrieve a list of article titles in which that word appears, pick the most likely, and start your reading. In any of the articles, you may see links or cross references to other articles in the same work, or even to other works. Perhaps in your quest for information about hypertext, you would see the phrases COMPUTER-ASSISTED INSTRUCTION or THEODOR NELSON. If the link is internal, you can follow it immediately. If it is to a co-packaged work (as an encyclopedia combined with a dictionary or atlas), linkage is still quite fast. If the reference is to another work altogether, not co-packaged with the one you are reading, there is no standard for how long it should take to switch over. The Internet is making such connections ever easier, but not yet universal, and some days the delays are terrible.

The first electronic encyclopedia I ever saw was not well designed and my reaction after about 10 minutes was to wonder why anyone who had a choice would choose this over the printed version. No more do I feel that way. The new ones are generally quite good and I like using them. The ability to cross-reference with other reference books is especially handy.

Another aspect of some of the electronic encyclopedias is that the user can ask a question instead of making a statement of interest or a search request. To an experienced computer-using adult, this should make little difference. But to a child it may be an important difference. The child can ask, why is the sky blue. The adult may know enough to simply ask for information about *sky* and *blue* or *color*. Technically, there is little difference because the computer is programmed to ignore the *why* and look for the occurrence of the substantive words of the question. But it can be more fun to ask questions and it moves us a bit closer to the computer HAL in the movie *2001: A Space Odyssey* or the intelligent talking automobile in TV's *Knight Rider*. It may make children more comfortable with computers. It may also give them unrealistic expectations.

Because the computer does not really understand the question, it uses the words of the question to find articles containing the same words. It really does not use the *meaning* of the question to find *meaning* in the articles, although there is research being done toward this end.[10] All sorts of variations are possible. The computer can look your words up in a thesaurus and recover some synonymous, or nearly synonymous, ones. It can remove suffixes so that *blue skies* matches *blue sky*. It can find words that often co-occur with your words. In the computer world, *artificial* and *intelligence* often co-occur. If you search for either, you might like to retrieve articles containing the other.

At our current stage of development, though, most commercial search systems could not generalize from *blue* to *color*. So, if you ask about the color of the sky and the article is about why it appears blue, you might miss it. On the other hand, there is research work being done on what is called *vocabulary switching*, of which this sky question is an example. If I ask for *blue* a vocabulary switching system could retrieve from its files the information that blue is a color, and so add the word *color* to the question. *WordNet* is a special form of dictionary or thesaurus that provides not definitions of words but lists of other words related in various ways to an entry.[11] Such a tool is invaluable in vocabulary switching. A common problem in using most retrieval systems would be met in a search for information about Theodor Nelson. Mr. Nelson's first name has no terminal *e* and should you use one in a search in most retrieval systems, you would likely get no hits unless you specifically remembered to instruct the computer to ignore any characters in the first name following THEODOR.

Having to sit before a computer, even a laptop, in order to use a reference book may be something of a put-off, but we are talking about heavy, multivolume books that, when used in printed form in the library, may involve walking back and forth from shelves to a table with each volume, eventually discouraging some interest in following up links. Other considerations are that although discs usually cost less than books, only one person at a time can use a disc unless a multiuser fee is paid to the publisher. Yet another factor is that, especially for high school students, the library is a social as well as an academic institution.

There is an intellectual advantage to hypertext, as well as a purely work-saving one. Give a person complete freedom to explore a subject without physical barriers to doing so — that is, allow the user to follow links at will — and that person can quickly build up a comprehensive mental model of the subject and its relationship to other subjects. This degree of comprehension does not follow from reading a single encyclopedia article however well written. I think what it means is that the agile, inquisitive, energetic mind has much to gain from the rapid retrieval of linked items. These qualities can be amply rewarded through electronic reference works.

Nelson envisions a world in which just about all literature is available to just about everyone.[12] He is not blind to the necessity of protecting copyright and the need for the people who build the networks and create sets of information (documents) to derive some income and have their work protected. Payments would be automatically authorized if copying were done. If an author created a new version of his or her own work it would be a new document, not a revision of the old, so readers could compare them if they wished. He uses the term *transclusion* to refer to the near equivalent of copying some text from one document into another. Instead of actually copying it, the new author would refer to the appropriate portion of the original. Readers could quickly "go" to the original, but there would be no actual lifting of copy.

Roger Blumberg describes a biology teaching program he created using the World Wide Web.[13] He enables students to browse through the original works of the geneticist Gregor Mendel, as well as other original scientific works, and to "visit" various laboratories and discussions. The key point is that students, even early in their study of biology, deal with original works and actual science. They *participate* from the start.

In fiction there is more controversy. The favorable aspects are that the reader of a novel presented in hypertext form can read the story from different perspectives. Read the segments of the novel that are directly about one character and this can give the entire story from that fictional character's point of view. Read a different set of segments about, say, a house, and you could get a narrative of how that building has fit into the lives of the characters who lived in it. Thus, the reader may find quite different stories depending on the self-chosen path. One critic, Jay Bolter, reports his experience with *Afternoon, a Story*,[14] so far the best-known and most widely reviewed hypertext novel:

> *Afternoon* ... is the sum of all possible readings, the sum of all the paths the reader can take in exploring the space of the text.
>
> I often returned to the same episode — sometimes intentionally, sometimes in spite of my best efforts — and yet at each return the episode was different....[15]

The reader of a hypertext novel plays a significant role in deciding what the novel is. This fits well with McLuhan's aphorism that "the user is the content."[16] Even with conventional text (or graphics, music, cinema, and so on) each work may appear different to each user and it is the user's interpretation of what is being seen, heard, or read that determines the content for that person. Of course, an editor or cinema or stage director also plays a role in deciding what a work of fiction is. But this is a middle role between author and user. With hypertext, it is the reader making the decisions at the time of reading. This makes the reader part of the creative process — to some a source of great satisfaction. It can give a deep insight into the interplay among characters and events, more than might be possible from following a set sequence, especially if the reader is not concentrating on the sequence as well as the content.

Disadvantages

We have to remember how new the hypertext novel is as an art form. No adult has grown up with it as even a culturally important form, much less the dominant one of the age. No writer has spent his or her life writing in this mode. While there are criticisms, to be fair we should remember that we are criticizing a new art form very early in its life and many of the objectionable features may disappear as readers get more used to it and writers and technology get better.

For nonfiction, critics reverse the main positive point. The ease of skipping from node to node could lead to or encourage superficiality. Instead of reading and absorbing material, users could keep moving about until they find exactly what they are looking for, or think they should be looking for, possibly having missed important correlative material upon which to base true understanding. This leads to the vision of a future race of children, growing into adults, whose approach to research even at the secondary-school level or higher is to find and combine snippets of other people's work, not adding any insight or interpretation of their own.

As an instructor and frequent evaluator of student papers in postgraduate education in the United States, Canada, and Jamaica, I have found this last point to be the single hardest aspect of academic life for beginning graduate students to adjust to. In secondary school and even many universities, all that is required of a student paper is finding some source material, extracting or paraphrasing relevant portions, organizing it, and duly acknowledging this as the work of others. The students do not add their own opinions or interpretations. More than once I have been told by students that they had been told by other teachers *not* to include their own opinions, but to report only what others had said. Nothing of themselves. How does one become an original and critical scholar this way? We can hardly blame hyptertext for this phenomenon, but we can be leery of its possible encouragement of lazy and unoriginal scholarship.

In fiction, many readers, including this one, find hypertext disappointing. I *like* allowing an author to set the sequence. I do not think of myself as lazy, but I read fiction for enjoyment or enlightenment. I read a text or professional book for enlightenment, to enlarge my knowledge base, or to find specific facts. When I find a novel whose sequence is disturbing, I may not continue to read it. The hypertext aficionados would say that is exactly the reason why I should be reading in hypertext, but I feel bad writing is bad writing. A work that is not in a style I like is not going to seem any better because I read in a different order. My reaction to the Joyce Carol Oates novel mentioned earlier (a hypertext link, just there) was initially a puzzlement, but gradually I realized there was a sort of rhythm to it, that a skilled author could jump all over the story and take me with her. And then I can admire that skill while I am reading the content.

If I had to find my own way through the material in Oates's novel, I might not do a good job of it. Further there might be some question about whether the work I do read if I do skip around is the work the author wrote. Is it fair to the author to judge her on the basis of how I chose to read her book, rather than how she intended me to read it? Would it be fair or even meaningful to publish a review based on the critic's reading rather than the author's writing? Whose novel would be getting reviewed? Possibly, hypertext fiction appeals more to professional critics and literary scholars than to just plain readers. The former may be much more interested in a detailed look at *how* the author makes characters develop and interact, while the latter are content to watch, enjoy, and learn from the interaction. To be fair, we might consider whether a work should be judged on its impact on readers, however they choose to read it, rather than only on readers who insist on following the author's sequence.

Another aspect of hypertext I do not like, which may be resolved in time, is having to sit before a computer to read. Even a laptop of today is not well-suited to reading for relaxation or lengthy study. An aspect of this is having to put up with the trappings of Windows™ while I read.

The trappings are boxes drawn around the text with menu options, some of which do not directly relate to the book I am reading. I appreciate good book and type design, and this ain't it. The publisher of one such work told me the text was originally written for a Macintosh, which would not have cluttered my screen so. Nonetheless, it was offered for sale in Windows form. It does not *have* to be this way, even in Windows, but there were time and cost considerations. Publishers have to get to the point where they will not release a book with such flaws. My copy of the CD-ROM version of Stephen Hawking's *A Brief History of Time* had some text and illustrations missing. This was seemingly a common error of the type often made by computer programmers (anyway *this* former programmer) but, again, I was surprised that a respected publisher released the work without having done more thorough editing. Yes, I have seen this happen with a printed book, more than once. The difference is that, with the disc, I am watching and listening and suddenly most of the screen goes black and the sound cuts off. That is more jarring to the reader than finding some of the pages of a book misassembled.

Directions

After a generation of virtual stagnation, hypertext and its close relative, computer-assisted instruction, have taken off. The reference works that now are being produced were technologically impractical in the 1960s. There were no personal computers. Disc memories existed but were physically much larger than today's and had far less capacity. So it is not all that surprising that not much happened to hypertext in the 1960s. It was a software invention before the existence of the hardware needed to really make it work.

About 1969 or 1970 I was part of a team that reviewed the use of CAI at two American universities. At one, we saw some film strips for use in a physics course, with selection of frames controlled by a computer. These films, not just text or still illustrations, made up a large part of the instructional material. They seemed to beautifully illustrate basic concepts of mechanics. At the other school we saw an economics teaching program that, instead of just presenting a graph illustrating the relationship between two variables, allowed the student to supply parameter values and then see the curve representing the relationship being built up over simulated time. This would allow a student to compare the effect over time of varying interest rates for a given set of other economic conditions. The economist member of our team was ecstatic with this manner of showing and letting the student explore complex relationships instead of simply hearing a lecturer assert them or reading them in a text. Another course was in elementary Russian. I tried it for a while thinking that in 10 minutes or so I should learn *something*, maybe just part of the Cyrillic alphabet. I didn't. The course seemed to lack structure and there was no meaningful interaction between student and mechanical teacher. Back on the plus side, multimedia CAI is being used to teach high school students the dissection of frogs,[17] which can save money, time, and squeamish feelings with a simulation of the real thing. Recently, also, a medical school has stopped using human cadavers in its basic anatomy training — too expensive. Instead, a multimedia computer simulation of a human body is used, just as with the frogs. Hypertext-CAI offer us the possibility of making more instruction conform to the aphorism

Tell me and I'll forget,
Show me, and I may not remember,
Involve me and I'll understand.[18]

Another aspect of the early CAI courses was that, however complex the subject matter, they tended to call for multiple choice or simple fill-in-the-blanks answers, which did not always fit the material well. The key to the potential of CAI is that students are frequently tested — a short question immediately following a unit of instruction — and then the course branches to another unit based upon the answer to that question or the history of answers throughout the course. This branching is what we now call linking. It is unlikely, then, that any two students would follow exactly the same sequence through the course. Sound familiar? I became convinced that the reason CAI never developed well was that the question-answer sets were simplistic. They were easy to write, easy to test, but not necessarily an easy medium for the student to express his or her understanding of, or to ask a question about the subject, and they were *borrrring*. Even book readers can ask questions, in a way, by use of the index and table of contents.

The modern electronic encyclopedia's ability to appear to answer questions bodes well for the future. While computers are not yet highly adept at understanding questions, they do keep getting better, and I can conceive that some day a computer could allow me to say that I am interested in the relationship between A and B and have it recover some meaningful material, even if these two concepts do not exist together in any one text. I do not expect the machine to understand the relationship, but to be able to detect that there is a relationship because of common use of words or symbols or joint references to A and B in the literature.

The future of hypertext, including its instructional variations, depends to a very large extent on authors' skill. The mechanical tools — good quality displays, sound, large memories, sophisticated language-analyzing programs — are now available. Let us see what the authors can make of them.

Students have a role to play, too, if hypertext is to become successful. I believe there is some new learning or relearning to be done. It is not necessarily their duty to take all the initiative, but one way or another, they will have to become accustomed to a new way of studying — probing, searching, asking questions — rather than passively accepting instruction.

Serendipity is the finding of unexpected or unsought-for treasures. It is often mentioned as one of the benefits of the old-fashioned way of looking for books in a library or just thumbing through a dictionary or encyclopedia. And there is a danger of losing it when searching with a computer.[19] Who has not had the experience of chancing upon a word other than the one being looked up, and saying, "Oh, so *that's* what that word means" or "Oh, how interesting"? The same thing can happen while walking through the stacks in a library on the way to the location of a specific book, or while browsing the table of contents of a magazine or professional journal.

If finding and viewing information through a computer, whether or not hypertext is involved, means always giving the user what was asked for and only what was asked for, will we lose these delightful accidents? The answer could be yes or no. If system designers concentrate too much on giving users what they asked for, our information systems could become too literal. But systems could be designed so as always to include some information *not* requested. Of course, the computer could not guess what extras might excite you. Some people have suggested storing an interest profile of a computer user in the machine. Then, anytime it is doing a search, it could tell you the latest news on that company you were interested in last week. We could go on and on, proposing different methods of adding random information to your output. The problem is that the more information the computer has about you and the more it uses that information to find unrequested records in its memory, the less truly spontaneous is the discovery. Can there be serendipity without spontaneity? And how important is it?

The hardware, which was wholly inadequate in the 1960s, is still not what is needed to make the new forms of publication take off. And I continue not to understand why publishers' representatives are not permanently camped in the offices of IBM, Apple, COMPAQ, et al., begging, cajoling, threatening, or whatever it takes to get these people to make good book-reading machines. Of course they could do it. They just do not seem to realize the potential. And the publishers and educators do not seem to be enlightening them. Why do the publishers not subsidize the development, before the public gets it into their collective head that a book on disc implies a book that is hard to read?

Now, to return to what I think is the critical problem. What would be the effect on thinking and learning of a massive switch of instructional and reference works and fiction into the hypertext mode? There have been some studies, but relatively few, and at this time I think we have to say that research is not yet giving us the answer. It is, of course, a very difficult question. It is always very difficult to test an educational technique and somehow control for the influences of teacher skill and motivation, content of educational material as distinct from the medium of publication, and the motivation of students. Trying to remove the effects of motivation from an experiment involving phenomena as complex as learning is a real challenge.

To go back to the pro and con mode, what favors the increased use of hypertext are: The cost of production and distribution via disc or telecommunication is probably much less than for paper publishing. Imaginative, motivated students might do better because they can so quickly find links from one concept to another. They can ask questions and, if the questions are imaginative and the software good enough this can further accelerate learning. This is the kind of usage Nelson envisions and for people as bright as Nelson it is, indeed, attractive.

I have not mentioned subject matter. Early CAI was assumed better for instruction involving memorization such as language learning or beginning biology and task performance, such as solving algebraic equations. It was clearly less adept at philosophy where the short, multiple choice answer was less likely to enable a student to demonstrate comprehension or the lack of it. As the computer's language comprehending skills increase (not part of traditional hypertext), the potential even for teaching such subjects should increase.

The other side of this coin includes several considerations. The bright students with lots of initiative should do well with hypertext material. What about the average or dull ones? Proponents of CAI have always pointed out that one advantage of that mode is that the computer is an infinitely patient instructor. I have seen a child with a learning impairment do beautifully on almost any computer task, working well ahead of the grade level of his other schoolwork, so there must be something to this claim. On what must be the third side of the coin, students with difficulties need patience, for certain, but they also need perceptive teachers and I remain unconvinced that perceptiveness of individual accomplishments or difficulties is truly present in the CAI or hypertext material that I have seen.

Another negative is that while a CD-ROM or Internet-transmitted book, article, or other document can cost considerably less than a printed equivalent, we must also consider the cost of the computers needed to read them. Hardware prices keep going down per unit of capacity, but today anything like an all-computer school and library would surely cost more than the traditional version. Further, the all-computer version would require a large expense in a short time to convert. Then there is the cost of maintenance and of the people who do the maintenance and the incessant demand to upgrade hardware and software. School boards love to provide schools with computers. It does seem to help get the members reelected. But school boards do not like the cost of complete conversion or of having to provide enough equipment that any student can get access to a computer when needed. Then there is the question of homework. Of course, lots of children will have computers in the home, but not all. How will these latter do their homework?

On which side of the coin does serendipity or its lack fall? I'm not sure. Its complete loss would be, to me, a strong negative. Can this random activity be reinserted satisfactorily? We do not know. We will have to see what systems developers have in store for us.

The potential for this new form of publication seems very great. The cost of development is very large. And the conversion cannot come rapidly, no matter how much money is spent. It takes time for all of us to adapt to the new technology: students, parents, teachers, authors, general readers, and publishers.

As this new medium matures, or tries to, we will continue to see conflicts about it. Some day, it will either have become an accepted and appreciated art form or it will have faded from memory in the role of a medium for fiction. It is too soon to tell. In the nonfiction categories, though, its success seems assured.

Chapter 7
Interacting with Information Machines

Interaction means two-way communication. It means two parties not only communicating with each other but having an effect on each other, so that an action or utterance of one party may depend on previous responses by the other. It is hard to draw an exact boundary between interactive and one-way communication, especially between a person and a machine, because real masters of fly rods, fountain pens, or sailboats will say that they can feel what the machine (yes, they're all machines) is "saying." The novelist Jan de Hartog put it this way, describing the feeling of a ship's captain as he mastered a ship newly placed in his charge:

> I was at last, in my turn, imbuing a piece of man-made machinery with a personality of its own, linked to it by a bond of loyalty and devotion, an umbilical cord of love.[1]

Aristotle told the famous parable of the man chained within a cave who could see the outside world only through shadows cast onto the cave walls. His point was to illustrate how our understanding of the world about us depends not on how the world actually *is,* but on our *percep-*

tions of it. This was a pure case of noninteractive communication. The man chained within could see transformed images, but did not know they were transformed from three to two dimensions, hence could not know what their true form was, and could not do anything to make those shadows take note of him. It also brings into question what we mean by "true," since the chained man's view is as true for him as anyone else's view is for himself or herself. In the early days of computing, many people felt that they did not know what a computer was or what it meant to them. Here was a new kind of machine, supposed to do wonders for us, but most people did not know what a computer really was or what it was capable of doing. Even after one was installed and operating, it was often unclear how to deal with it. It seemed intelligent, on one hand, but showed no sign of listening to us users say what we wanted it to do. Only a priestly caste called programmers could talk to it.

The Nature of Interactive Machines

In the 1970s, the term *interactive television* was in vogue, more as a topic of conversation than as a reality.[2] Broadcast television, by its nature, is not directly responsive to its viewers. That's what broadcast means — the same message going out, one way, to multiple recipients. There is no corresponding mechanism for viewer-to-broadcaster communication. If there were, there would be the issue of which viewer or viewers to respond to, because broadcasters could hardly pay individual attention to millions of simultaneous incoming messages. And, if all this were resolved, how would the broadcasters or actors respond and react? Can they change who-done-it on the spur of the moment or call back an errant football pass and do a running play instead because that's what the audience wanted? A few experiments were done with panel shows, where audience consensus could guide the subject or approach the panel took, but this never came to anything. We now have a great many talk shows on both radio and television in which a host reacts to some individual in the studio or broadcast audience. Usually, it seems, the more outrageous the audience member the more likely he or she is to be heard. These are interactive, but most viewers do not participate in the interaction, and the interaction is really between people, not between the person and the TV or radio mechanism. All that is really happening is a televised or radioed version of a conversation between two people.

In sports television, where some of the most imaginative camera work is to be seen, we have two features that could evolve into interactive TV. There is the rerun. A play is viewed again and again, from different camera angles, giving the TV viewer often a far better picture of what happened in the game than the live audience gets. Live television is increasingly being presented on a large screen in the stadium or arena of professional games and those replays affect behavior of people in the arena audience and even the players and referees.

Then there is the camera attached to a sports participant. So far, this has appeared in auto racing, where we are occasionally given a view of the track and other cars as seen from a camera mounted inside one of the cars, and baseball where a camera has been mounted on the catcher's mask. In hockey and basketball they come close by affixing a camera to the goal, giving a view very close to what only a hockey goal tender could have seen previously and in basketball the camera is on top of the backboard, giving a view no one in the area has.

Now, there is talk of putting a miniature camera in the helmet of a hockey player. It might take a separate channel for each player so outfitted, but then the TV viewer could select the player he or she wanted to follow. You could see the game from the perspective of Wayne Gretsky, or of the goalie, or of the player who has to defend against Gretsky. The same could be done in football, and even in baseball the viewer could be given a choice of the camera through which he wishes to view the game. If cameras get small enough, how about one in the waistband of a pole vaulter, or perhaps the throat of a tennis racquet?

Actually, in all these cases, the viewer would be switching channels to see any one of the, say, ten simultaneous telecasts of the same game. But merely switching channels is a low level of interaction. It does not directly affect the conduct of the event unless coaches begin to respond to audience preference for which player or even which kind of playing the audience wants to see. Even without this last level of interaction, multiple cameras selectable under user control could radically change what it means to watch a televised sporting event.

In one sense, even the remote control device for a home television receiver provides a form of interaction. Using the remote, a watcher can quickly change channels, of course, but can also put all advertising on *mute*, or with a fast enough hand on the buttons, can watch two events nearly simultaneously. If the TV is equipped for inserting one channel's view within another, the fast operator can play with four simultaneous events, madly switching back and forth. If the ratings people who monitor a family's television watching see this sort of behavior on a large scale, it would have an effect on programming, but not an instantaneous effect.

All these variations on camera location would be a boon to televised sports, but it is still not interactive in the sense of the viewer having an immediate effect on the action. The viewer's impact in these multi-view options is on his own receiver, not the players (I say "his" because I am hearing, over and over, and observing personally, that this flipping from channel to channel is a "guy thing." Similar things could be done even with the news. A viewer might be able to "say" to a newscaster the equivalent of "Let me see more detail of that map" or "Who is this person, X, you're talking about?" Then, the answer could come back on another channel or a private telephone line, feeding into the computer for printing, delayed viewing, or whatever. It would be private because not every other viewer would want the same treatment. It would not have to be completely individualized. The broadcaster could prepare a dozen or so messages enlarging on various aspects of each news item. Or, in sports, where there might be as many backup channels as there are cameras in the arena for each primary channel, giving the viewer a choice of coverage of the story.

Books, today, are not interactive. They may trigger all sorts of reactions in our minds. We may be moved to laugh, cry, or shout at the book, but it is written and printed when we get it, and our reactions are not going to change that. Might they someday become interactive? Yes. Hypertext is a form of interactive book, although the text has still been written in advance. We cannot change it ... at present. Just ahead, we'll fantasize about a book we *can* change.

Interactive Computing

As I have mentioned, early computers were slow by today's standards, and terribly expensive. The hourly cost of operating the first computer I worked on was estimated at $300. I started out earning about $2 per hour (with my mathematics master's degree). So, in administering a computer center, the *machine's* time was protected, not that of the programmer or user. We wrote programs, submitted them to people whose job was operating, not programming, the computers, and waited for our allotted time when the program would be run, which might even have been the next day. The results were printed out. If there was an error (a programming error, data error, or if we accidentally asked for the wrong thing to be done), the instructions had to be changed and we ran another cycle, probably again involving a day's wait. If there had been a misunderstanding between the end user, the one who needed the output, and the programmer who created the computer software, this could take weeks to resolve, and led to a great deal of frustration with computers.

In the 1960s, computers changed from the mode just described to one involving *time-sharing*. This meant much the same as it does today in vacation real estate. Each user could get a slice of computer time, when needed. The difference though was that the computer's time slices were in terms of milliseconds, switching from one user to another so fast most people barely noticed that anyone else was "on." Each user was able to feel that he or she was in complete charge, but the cost of using a computer in this way was much less than if the computer had really been totally under one person's control. This is because a user pays only for time actually used, not time spent waiting a turn or just thinking what to do next. This mode enabled people to try ideas, get quick feedback from the computer and, perhaps, try something else, all without lengthy delays. But the computer was still under central administrative control, and at times a given user and the central administration were not in agreement over priorities, and this could mean slow service or nonavailability to some users.

When personal computers came along in the 1970s, they returned us to the old way of using a computer, putting it at the service of only one person at a time, but the cost was very low. That person now had full control over what programs were run and over any other priorities concerned with its operation. The same applies to small processors embedded in other machines, such as a fax (telling, for example, whether an incoming message is from another fax or from a voice telephone, and treating it accordingly) or a VCR, allowing for user programming of recording operations. But the level of programming in PCs allows us to enter some data, run a program, see the results, make a change, run it again, all within a few minutes or even seconds, and at a reasonable price. It allows the machine to remember the kind of choices we made and automatically make them for us next time, as some automobiles remember how we set the seat adjustments last time. It has become like talking with an intelligent assistant. Try this. What did it give us? Why? Try something else. What did it give us? If I make an outright mistake the computer will point it out to me and tell me how to correct it.

In my own specialty of information retrieval, we have seen great strides in interaction. I ask a question: FUND INFORMATION ABOUT X. The computer tells me that FUND is not a recognized command. If it's really smart it will ask if I mean FIND, the command closest in spelling to what I asked for. I get that straightened out and then review some retrieved records, perhaps stories from a newspaper file. They're not quite what I wanted, but I can see why. I didn't explain it well, or it didn't understand subject X well enough. This happens often enough in real life. I rephrase my request. The results are better. I see some terms used in the articles that I didn't think to use, and decide I might search for other articles containing them. I do, and the results get even better.

Another way to do retrieval is to allow me to enter a natural language statement of what I want. No commands, just a description of the subject matter I'm searching for. The computer, as we noted in describing how modern encyclopedias operate, looks for articles containing the same words I used, or words similar to mine. This way, the user does not have to be concerned with remembering the specific commands, or to encounter the frustration of making an error when typing a command, as I simulated doing above.

Interactive computers let us do analogous things with, say, architectural drawings. Make a sketch, look at it, ask colleagues to comment. Keep changing it. Go back to an earlier design when you want. Have the computer calculate square footage, discover you've exceeded some zoning restriction, make the necessary change — all this in a very short period of time. Then, when the design looks final, the architect could still get quality drawings on paper in just a few minutes.

Generally, people who have experienced this kind of computing will never go back to the old kind, where seeing the results of what you asked for could take a day or so and any error would mean a wasted day. But, it's not just the speed. It's the sense of working with a partner, or getting immediate reaction to something you wanted to try, that is actually exhilarating. One of the earliest people to recognize the value of this kind of interactive computing was J.C.R. Licklider, who was then doing a lot of calculations associated with acoustical design. This involved having an idea, doing many calculations to compute the effect of the design, possibly finding defects, and changing and trying again. In 1966 he described what he called a symbiotic relationship between computers and their human users. *Symbiosis*, in biology, denotes two dissimilar organisms living together in interdependence, to their mutual advantage. Licklider put it this way:

> [M]any problems that can be thought through in advance are very difficult to think through in advance. They would be easier to solve, and they could be solved faster, through an intuitively guided trial-and-error processing in which the computer cooperated, turning up flaws in the reasoning or revealing unexpected turns in the solution.... One of the main aims of man-computer symbiosis is to bring the computing machine effectively into the formulative parts of technical problems.[3]

Although we may think of interactive computing as a very modern phenomenon, Licklider wrote in 1960, and I worked on an interactive system in 1955. The big difference between those early days and now is the PC. The price of a computer is now within reach of ordinary people, and control of the machine is entirely in the hands of the owner or user. Mind, I did not say computers are within reach of everyone. But in 1955 the idea of a personally owned computer was not even considered. By 1965 there were a few people working on the idea. By 1975 a few people had them, and in 1996 some 30-40% of North American homes have them. So, the direction is obvious. Every year they become available to more people.

Virtual Reality

Actually, I hate this term, but it has become too popular to look for a replacement. It is an exotic form of simulation, which as I pointed out in Chapter 1, has been with us for some time. And while I don't like the expression, the reality it represents can be very exciting. If you are interested in reading this book, you have probably had the experience of interacting with a computer but probably you have never had the experience of dealing with a noninteractive one. Let us take a giant leap forward, beyond simple interaction. The way you talk to a computer normally is primarily by use of the keyboard and the mouse, that little pointing gadget that gives you a sore wrist. There are computers that read handwriting and there are those that can recognize words spoken by a human voice. Both require a lot of adjusting, because we humans can speak and write the same words in wondrously different ways. As I write, these modes are not quite in the mainstream of computing ... yet.

Virtual reality gives you more vivid displays and several different ways of sending messages to the computer.[4] If you were using a VR system, you would be wearing a helmet with some odd-looking goggles attached. The goggles let you view a three-dimensional world, projected by the computer through the goggles. My limited experience as a user showed me a rather fuzzy 3-D world, which is also what I remember of 3-D movies that came out in the 1950s. As a VR user, you wear an instrumented glove or hold some form of instrument in your hand. With this you can point, simulate grasping, and then move or turn some simulated object that you can "see" through your goggles. Stop here and you have a system that could be used to train people to handle dangerous materials or instruments, and even, if the images were sharper, to train surgeons or art restorers.

But VR goes yet another step. There is a small transmitter and an antenna in your helmet and sensors around the room pick up signals that tell where you are and which way you are facing. Similar transmitters could be placed on your hands and feet. Now, we could train western gunfighters, real policemen who must face armed adversaries, or other people having to make very complex motions possibly coupled with important decisions in little time. Now we could let a person dance and record the movements well enough to instruct another person how to dance the same way. We could simulate a football game where you are the ball carrier, snaking your way through hostile linebackers. We could simulate real baseball pitchers and give batters a realistic sense of what it is like to bat against them.

Yes, this takes a lot of computer power and, except for the Nintendo kind of computers, VR is too expensive and imperfect for many real applications ... yet. But, when engineers get this far, it becomes only a matter of time before they improve the machines and bring the price down. What has all this got to do with books? Let's consider.

Virtual Fiction?

We have, in effect, four forms in which fiction is presented today: live (unrecorded) theater including storytelling; cinema or television (recorded theater); conventional printed stories and novels; and hypertext. To me, multimedia computer presentations are a form of recorded theater, delivered not through a motion picture projector or television channel, but through your PC. We'll discuss multimedia in Chapter 8, but the point is that it *looks* like a movie to which hypertext controls may be added. The first three of these fiction-presenting forms are major art forms and they also represent major industries. That means that a great deal of attention and effort are expended on studying how to create the forms, studying what the public wants or will buy, producing them, and criticizing them. There have been a few experiments with multimedia fiction presentations with variable, reader-selected outcomes. It is not yet a significant factor and shows no sign of a significant level of investment ... yet.

Computer games have been popular for nearly two decades. They may be seen as a form of fiction, but there is no predetermined story line and mostly what you get out of them is the sense of being in something of an athletic contest or in combat. Situations are presented, the player makes decisions and programmed consequences follow. Emotions (except maybe fear or competitiveness) or philosophical positions don't enter into them. Would it be possible to "write" a real story such that the "reader" is a participant, as in a game, but where the game is something more than finding the treasure and avoiding being beheaded by the ogre? The situation would have to be like a real novel — a story with interaction among characters, not just simulated physical actions.

What would have to happen to do this? First, the decisions faced would have to be posed within the story, not presented as a break in the narrative to ask the question "What do we do now?" The smoother the integration of the question or decision alternatives into the story, the more real the experience is going to be.

Second, if the hero sees something untoward happening, say in an apartment across the way, as in Hitchcock's movie *Rear Window*, he can decide to call for help, continue to watch, or do nothing. In the movie the hero was relatively immobile due to an injury, hence unable to take direct action himself, a clever ploy to limit the dramatic possibilities open to him. If the reader as protagonist does nothing, there isn't much of a story, so the author would have to prepare some plotline that would, eventually, force the person to become involved. There could only be a modest number of prepared alternatives and, again, the smoothness of their integration into the story would determine the sense of reality the reader experienced. This would be especially true if the reader tried to do something unanticipated, that is, something the author has not explicitly prepared for. So, there has to be an 'all other" contingency and it has to seem real enough that the reader segues into it smoothly. Yes, all this calls for a great deal of authorial skill and no author is going to acquire it overnight.

Couple these possibilities with the ever-improving ability of a computer to understand speech, and we could have a novel in which the reader *is* the main character, observes scenes, hears voices of other characters, and can talk to them and be understood and reacted to — interactive computing. Couple these capabilities with virtual reality and we could have the reader-hero slam doors, walk out throw things, or shoot at people. Sorry about the violence of these actions, but violence happens a great deal in fiction. Another thing that appears in fiction a great deal is love, both the emotion and the action. The emotion would pretty much have to be expressed in words or gestures. The experience of action I leave to the reader's imagination, but remember, we would have both VR and inflatable dolls, at hand so to speak.

True, a potential buyer of one of these works would have to select carefully. If you don't like shooting, don't buy a hard-boiled private detective story. If you don't want to play the part of the male lover in a heterosexual romance, don't buy the work. By the way, a good bit of modern communication technology first achieved popularity with sexually oriented or pornographic material. X-rated videotapes were big sellers before conventional movies moved into the market on a large scale. On computer bulletin boards and electronic mail, sexually oriented materials have often accounted for a large part of the traffic and I have heard it estimated that about one-third of questions posed to some of the Internet information retrieval systems include the word *sex*. It would appear that there is an unfulfilled need out there and any new technology that helps to satisfy it becomes quickly popular. One effect of this is to pump needed development money into the communications technology industry, providing experience for creators and producers, and paving the way for future main stream productions. So, if my vision of virtual novels is to come about, it will probably start with the raunchiest material and may take a generation to become a generally accepted art form. Look for some of the first VR for adults to be of the porn genre.

Is VR a Good Thing?

There are three easy answers to this question: yes, maybe, and no. The choice depends on the circumstances of use.

On the "yes" side, there are some very obvious beneficial uses of VR, largely in training, especially when either the task or the tools are expensive or dangerous. People can be trained to handle highly toxic substances, to dissect frogs, to do space walks, or to "see" inside and "navigate" through a complex structure such as a large organic molecule. *Training* tends to mean teaching someone to *do* something, rather than to understand or to feel (in the emotional sense). The line is not clearly drawn. The essential point is that VR has many valuable, positive uses or potential uses.

On the "maybe" side lies education involving complex, nonquantitative concepts. Can VR help one understand different perceptions of God among different religions? I tend to doubt it, but then good filmmakers can do most interesting things by way of graphic images to get across complex ideas. So, I don't want to prejudge this point.

No! Do not use a machine to interact emotionally with a person, especially a young and impressionable one. We have Microsoft telling us about the wonders of their new storytelling CD-ROM,[5] intended to do at least part of the job that children's librarians, teachers, or parents have traditionally done. That sounds innocent, even beneficial enough. But we see in science fiction persons interacting with humanoids or with creatures like Mr. Spock of *Star Trek* who are without human emotion. Is this good? What is *good*? As a person who has written interactive computer programs and diagnostic programs intended to serve the role of coach to a computer user, I am disturbed at the prospect of children growing up with computers as their friends. I want the computer to remain a tool, unique and valuable, but not human. I want the storyteller to be human.

I have to confess that when I think of the boundary between a computer seeming to be the narrator and a motion picture showing narration, or a cassette player playing back a recorded narration, I don't know why I object so strongly to computer storytelling. Maybe it's that I want the hearer or viewer to be sure he or she knows that what is happening is the playback of a recording made by a human being. When, or if, the audience thinks it is the machine telling the story, I worry. And, back to the other side again, what small child hasn't at some time thought the TV images or radio voices were made by little people inside the box? I don't know of anyone who did not easily shed this idea with maturity, or to whom holding this view at one time was any sort of impediment to later learning or to forming healthy human relationships. Before I get too negative, I have to remind myself how much *I* like books on tape, how well-produced stories might attract children to literature, and that because they might appreciate their literature in a different medium than I did does not make them illiterate.

The Virtual Imagination

This section was originally written as a spoof, but some concepts are not spoofable. It has some potentially serious content.

Only a few classes of people can have avoided hearing about virtual reality. They are likely to devote their reading to sports or politics. All others, including stock market mavens, know that virtual reality, or VR, is the greatest achievement in computer science for the benefit of humankind since the last greatest achievement in computer science.

What is it and why is it so valuable? Well, a person participating in VR (a virtual realizer?) wears a bicycle helmet to which is attached a large set of goggles, making him or her look like a bug ready for a road workout. The realizer holds in one hand a pistol-grip device with some buttons on it. What you see through the goggles is a virtual world created by a computer. It might be a building (well, a virtual building) representing what an architect is thinking of designing if the client, upon virtually walking through it, approves. The client uses movements of the body as well as the buttons on the handheld device to maneuver through the house. You can start in the entrance foyer, turn around, see a door, and propel yourself through it. You can stop to admire the art on the walls or the view through a window. You might even jump out the window to test safety features or suicide adaptability of the house. Got the picture? (That's an intentional pun. We say such things in conversation frequently, hoping that we have, in fact, conveyed an appropriate image in the mind of the hearer. Since we cannot physically create the image in another person's mind, we can only ask if the person feels he or she has one.)

Another application would be to fly an airplane without, of course, leaving the ground or even your air-conditioned lab. What's that, you say? People have been simulating the flying of airplanes since the 1940s? Well, yes, but it wasn't called virtual reality then and, you see, it's the *name* that is the real accomplishment, the real advance in knowledge. What it does is create other possibilities, and gives us a chance to recognize some other categories of existence we have long had but not recognized for what they were.

Virtual reality, as you may have noticed, consists of two words. The second describes a category of existence — reality — something real that exists more or less as we see it. *Virtual* is borrowed from optics. A virtual image is one where you see something that looks real, but what you're seeing isn't there, or anyway isn't where you're seeing it. Look in a mirror. That scowling face isn't there in the mirror. It's outside the mirror looking at its virtual self.

If we can have virtual reality, what about the other linguistic combinations, shown in Figure 6. How about actual reality, or as the physicists would say, real reality? Well, we have plenty of that. Maybe we can simulate flying an airplane, but when a pilot puts one into the side of a mountain, that's very real. There are mystics who will say nothing is real, that all is virtual, and there are even physicists who will say that while that airplane crash was real in this world, there is another world in which the airplane didn't crash. So, to have real reality, you have to be willing to believe that there is reality in the first place, and if there isn't, then all is virtual. See what VR has done to our linguistic ability? We can talk on a higher plane now.

Let's skip over virtual unreality for a moment, because that's where the real fun is. (Or is it virtual fun? See what I mean? The language aspects are far more fun than the VR.) Consider real unreality. What's that? Well, actually, it's a novel, or a fiction movie. Or it's the *Odyssey*. It's something we've had for years, but didn't know what it was until computer science opened our eyes. The book is real. The story portrayed is not.

Now to virtual unreality. This means something that only seems to be unreal. It can't really be unreal or it would be real unreality. So it has to be real, but appear to be unreal. It is art imitating life. It is the O.J. Simpson trial seeming more like a soap opera than an actual trial of a real man for a real crime.

Summary

It seems clear to me that the electronics world has really come up with something important in interactive machines. Yes, from one point of view,

	Real	Virtual
Reality	Airplane crashing into mountain side	Simulated plane appears to crash but no actual crash occurs
Unreality	Novels, TV fiction Motion pictures	Art imitating life. Real events made to seem unreal

Figure 6. Variations on virtual reality. *Taking the adjectives* real *and* virtual *and the conditions* reality *and* unreality, *we have the combinations shown here. All these forms exist today.*

we've had them for years, but interactive computers and televisions give the average person control without having to become an expert. People who deal with machines, whether as computer programmer, VCR user, or bank ATM user, like to feel that they control the machine, not vice versa. It is a frustrating and demeaning feeling to face a machine that will not do what you think it is supposed to do for you. But, when a machine *does* do what you tell it, when it seems to talk to you if it doesn't understand, when it says the equivalent of "Yes, sir" or "yes, madam" when it does understand, we seem to like that. Look for more in all sorts of machinery in the future.

Chapter 8
Multimedia

Remember that multimedia books are not new in concept if we take the broadest meaning of *multimedia*. The first combination of text and pictures came quickly after the European invention of movable type. Motion pictures with sound are multimedia. Very few people can still remember the pre-"talky" era of cinema. Other combinations include the scratch-and-sniff inserts in magazines, usually selling perfume, and the Christmas cards that play a tune when opened. What is new is the number of media in use and particularly the use of the computer as a base channel that carries the others. What is not new is the abundance of claims for the benefits that the new media will bring. Maybe they will, but the extravagant claims have been heard before. Here is a quote attributed to Thomas Edison:

> I believe that the motion picture is destined to revolutionize our educational system and that in a few years it will supplant largely, if not entirely, the use of textbooks.[1]

What Is or Are Multimedia? — A Review

In cases when books have been published in disc form, they almost invariably use a multiple media format. A few consist of just text in digital form or even in video form, i.e., pictures of text. But a CD-ROM can hold hundreds of conventional-sized, digitized novels. Few people would want to read one of them on a computer. CD-ROM users are likely to be analysts of language or writing style, or people looking up quotations. A book on disc, then, is a multimedia production, and this means we use at least two media, pictures and text. Commonly, they can and do also "display" some sound. The pictures can be stills or full motion video, and the sound is going to be far more than the beeps of a computer program telling us we have hit the wrong key.

If we are dealing with a typical reference work — encyclopedia or dictionary — there is mostly text with some illustrations and occasionally some sound. There may be short motion sequences. If it is the electronic equivalent of a coffee-table book, say about Beethoven, you are going to hear Beethoven's works played by a symphony orchestra. If it is about the art collection of a famous museum, you are going to see reproductions about as good as those in an art book, probably better than the typical 8-by-10-inch reproductions sold in many museum shops, but they may not match the color quality of a 35 mm slide made by a professional photographer. Typically, as in a motion picture, there will be full-color graphics always on view and almost always some kind of sound, whether background music, narration, or perhaps an example of a musical instrument playing a few notes to demonstrate its particular tone and range. But the media are not just for bringing information out of a computer. The keyboard and mouse are media for entering information into the computer. We also have scanners for entering graphic material and direct input of video images from a still or video camera, and to some extent we can write in script or speak to the machine. With virtual reality we add still more ways to send information both into and out from the computer. Multimedia works are usually in hypertext form, i.e., with the user having a great deal of control over the sequence of presentation. The technology of multimedia is complex and the manner of its use is demanding.[2]

Possibly the best analogy to describe multimedia works is that they are like motion pictures that may involve both photographs and animation, sound, and some text display *and* with the addition of some way for us in the audience to talk to and affect the sequence of the presentation. There is no fixed set of rules governing which media have to be used, or how. And since we have different words to describe our perception of different types of communication — reading, watching, listening, smelling, feeling — I am henceforth going to use the word *read* to mean any of these, including reading music or reading odors.

The Appeal

Why all the excitement? It's simple — people seem to like it. People liked illustrated books. To some readers there seems to be something daunting about books with text only. Even just moving from a single body of text to one subdivided into chapters seems to help. Beginners are thrilled to get their first "chapter book." Reading a book so divided seems to give a child a sense of being in an advanced state of ability to read, marking a rite of passage. But I can still remember browsing through books when I was young, before undertaking to read them, to see if there were any illustrations. I cannot remember the age at which I did this, or stopped doing it. It did not seem to occur to me that the text was the same, no harder, no easier, whether or not there were pictures. But, I was convinced I would enjoy it more with that extra channel of communication — the pictures.

Talking cinema took very little time to drive silents out of the market and color television, while it has not entirely done away with black-and-white receivers, has come close. These new developments were not fundamentally different media. They were combinations of existing media that viewers were used to, but not used to seeing or hearing together. It is interesting, though, that moving pictures have not done away with still photography and there is still some market for black-and-white cinema. Probably, it is because audiences recognize these as different media. There was something of a furor in the 1980s when colorizing of black-and-white films was introduced.[3] It seemed not to matter with grade B westerns, but coloring classics was something else again, and was much resented. There have been lawsuits over the issue, generally brought by artists rather than copyright owners, claiming their work is being harmed.

There have been studies suggesting that background music helps to hold audience attention on training or educational films. I am not sure if there is actual proof of this, but the practice is almost universal in such cinema. Certainly, we rarely have any significant time in a fiction or documentary film without some background sound. So the assumption is nearly universal in the film world that a second channel of communication helps the audience concentrate on, respond emotionally to, or appreciate the message carried in the first medium. Indeed, even in the silent film days there was usually a live piano or organ accompaniment to the film.

A computer–user interface is the set of programs and equipment that produces the combination of output displays and sounds the computer sends to the user and the language or methods the user employs to convey information to the computer. Some designers believe that there would be more effective communication with users if the computer talked to the user as well as presenting written messages. It surprises me, therefore, how rarely we see it done. Yes, we get those annoying beeps, but how about a voice saying, "You have asked to read a file that does not exist or is not stored in this computer. Would you like to try again or to look at a list of files that are stored here?" Just as some well-selected music helps us concentrate on understanding how the sea turtle manages to navigate across the ocean, a pleasant voice explaining the file name error and suggesting alternative actions might make the experience less frustrating. It can be overdone. It has always been my experience that overly cute messages from a computer are not appreciated and rebukes are very much resented. I hope this means that we users are hanging on to our feelings of human superiority over machines and that we might tolerate a teacher yelling at us occasionally, but not a machine. The recent win over the reigning world chess champion by the IBM chess-playing computer, Big Blue, may set back our own feelings of superiority.[4]

Microsoft Corporation calls much of their line of multimedia products *edutainment.* You can look at this two ways. The cynic might suggest they have caught on to the idea that if they slip a little learning into an entertainment work, parents will readily buy it. The other view is that Microsoft has learned that learning can be fun, that more learning takes place when the student is motivated, and that enjoyment is a great motivator. Such an attitude may offend some Victorian senses of propriety, but I think they are on to something meaningful.

A recent news item reported on a study of the effect of odor on behavior.[5] The scent of cinnamon buns seemed to outperform perfume in inducing sexual arousal of human males, but more importantly, the odor of strawberries helped people to perform better on educational tasks. A physiological explanation was offered: that the area of the brain that responds to olfactory stimulation is linked to that used for certain intellectual tasks. So, like background music, there is some indication that use of parallel channels of communication does improve human understanding of information.

Overall, then, the basic assumed benefit from multimedia, whether to enhance learning or enjoyment, is that if information comes to us over more than one sensory channel at a time, we pay more attention and we get more out of it. It need not be the same or similar information in each channel. Some signals serve only to enhance receptivity to or focus attention on others. An ever greater percentage of desktop and laptop computers is being sold with multimedia accoutrements — CD-ROM drive and stereo speakers. Probably, these will soon become standard in all computers larger than pocket size.

The Threat

Why use the heading *Threat* for something that seems so attractive and practically useful? There is concern that multimedia could displace text as the primary means of conveying recorded information among people. It seems safe to assume that if printed, largely textual books are going to be replaced, it will be multimedia productions that do it. The threat may be only to our current traditions, i.e., as youngsters become used to the idea that all media are multimedia, just reading text, looking at pictures in a gallery, or hearing music without pictorial accompaniment may be unacceptable or so unappealing that unimedia become ineffective means of communication. So if you're a little old-fashioned, as I sometimes am, this is a threat. If you are a pure modernist, then this is simply the reality of a changing style of communication, just as we have had other changes in the past. Let us consider first whether this change to all-multimedia *is* likely to occur and then whether it is a good thing or not.

How the Change Might Come About

We are already seeing electronic reference works displacing printed ones, and at least one encyclopedia publisher has predicted that such works have no future in the print world. Why? The multimedia ones are better in many ways and cheaper if the computer is already available. The biggest drawback to multimedia is in the realm of fiction. Having to read novels through the medium of a present-day computer is too much of a put-off.

We don't read encyclopedias for hours on end, nor do we necessarily expect to gain pleasure from reading them. We do expect to find nuggets of information and copy some text or pictures. Most electronic encyclopedias allow copying to be done easily, and that is almost benefit enough. Remember sitting in the library painfully taking notes on pages of text from library books? The photocopier did away with much of this, and the electronic book may finish the job. For better or worse, note-taking and photocopying will disappear from the scene. Encyclopedias are tools, and as such they work very well in electronic form.

Moving history books or essays into multimedia form presents a different challenge. These traditionally are long texts, possibly illustrated with maps or photos of the people, objects, places, or events described. If modern enough, sound would help. How different it must be to hear Roosevelt's, Churchill's, or King's actual words and intonations, rather than simply to read the text of their speeches.

Hypertext will have appeal in many book forms. Sequence is important in printed books. Yes, we may always want to jump around a bit or to look up some fact or quotation, but their basic intent is to be read from beginning to end.

Novels are being written in hypertext form, of course. I, personally, am not sure this is the mode that will survive. But I can conceive of print works gradually losing out to more cinematic presentations — multimedia.

Poetry is an enigma. It does not really contribute a great deal commercially to the publishing world, and is probably a net financial drain on it. But many publishers continue to produce poetry out of a combination of wanting to add to their cultural reputations and perhaps to do their bit to preserve an important aspect of our culture. Some poetry is being done in multimedia form.[6] One advantage is that the "reader" has a choice of reading the text or listening to the poet read it. The readings could be illustrated either with films of the reader or perhaps of the kinds of images the poem conveys. Is illustrated poetry the same medium as printed poetry? I think most poets would think not, so if the world were to go this way it would imply a change in their way of working or perhaps a change in who becomes a poet. The Canadian poet Phyllis Gotlieb told my electronic publishing class that she considers reading her work to a live audience a different medium than writing to them.

Magazines, journals, and newspapers vary considerably in their potential use of multimedia. Some, such as *National Geographic*, are almost saturated with high-quality, color illustrations, and people like to keep back issues in their own private archives. Daily newspapers are produced under severe time pressure and lack the paper quality for high-resolution color reproduction. They might like to switch to an electronic form wherein they could include more graphics and not be concerned with lasting power, except in libraries. Such a change is a big risk for them. Would their customers and advertisers still see them as the primary source for serious news? I think that issues of delivery time and cost will dominate what happens to this class of publication. Today, even on the Internet, some electronic newspapers avoid illustrations because of the cost and time required. But there is change. The *National Geographic* has published 108 years of their magazine on discs and even the venerable *New York Times* has begun using color photographs in its news sections.

Memory and computer speed are important factors in the use of multimedia computers. It takes a lot of computing power to display moving pictures. It also takes a lot of memory to store high-resolution graphic images, thirty of which must be displayed per second to appear to the viewer as real motion. Hence, in an encyclopedia, while there may be short, full-motion sequences, there are no long sequences. But the computer industry has always solved all problems of speed and capacity by more or less doubling capacity every year or two and occasionally coming up with a really big jump. Just coming to the market is the variously called digital video disc or digital versatile disc (DVD).[7] The capacity of this device is about ten to twenty times that of a CD-ROM, about 6 to 12 gigabytes, compared with 600 megabytes. A DVD could contain an entire full-length motion picture, something not possible with a CD-ROM. Used for the equivalent of a book, it could contain huge quantities of text and still images, as well as long sequences of motion video. This device could push the tape-driven VCR off the market.

The conclusion I come to is that many or most reference works will benefit from the multimedia form and will quickly convert. Journals and magazines may be next. Other types of works will not go so fast. Fiction has the stage, cinema, and television as alternative forms to printing and these, of course, have been with us for some time. Fiction writers who feel the urge to augment their works with visual images already have outlets. And, of course, we still lack a computer that will serve as a comfortable reading medium. So, progress (or perhaps I should say "change") in this genre will be slow. Change for other forms of the book will fall in between. I'm sure some historians will find multimedia irresistible, if they can find a market. Much will depend on how computer hardware develops. The increasing presence of computers in homes, schools, libraries, and offices makes dependence on them ever easier to accept.

Now to the question of whether we want these changes.

Desirability of Change

I see two basic aspects to this question of whether the change is desirable — the change being the shift from primarily printed books to primarily multimedia works. First, do today's kind of people want the change and would they benefit from it? Second, would the changes be acceptable and beneficial to the next or some subsequent generation?

My personal feelings and those of most of my contemporaries and also of students whom I have asked about this are that they like books. More than like, they cherish them. The disappearance of the printed book would be a disaster for them. It is also true, though, that most of the people I routinely deal with are computer users. Their uses are most likely word processing and electronic mail, but increasingly they are browsing on the Internet. The point is that they have become habitual, computer-dependent people. They are no longer technophobes. But they still do not want to do routine reading through the medium of a computer. These people, who are also book-buying people, will probably not allow the multimedia form to displace the paper form in the near future. Conceivably, rises in the price of paper could change that.

The second question has two parts: will multimedia forms be acceptable and will they be beneficial? The one thing I am sure of is that the acceptability of multimedia will increase with subsequent generations. How long it might take to see multimedia replace paper I cannot say. That my youngest child will have less objection than I have, or that his children will have even less, I am sure of.

Beneficial? I go back to some of the objections to hypertext and I acknowledge that we do not really know how this will work out. There remains the danger that as all literature shifts to high-intensity forms, requiring ever less intellectual effort or imagination on the part of the perceiver, concentration, creativity, and reasoning power will suffer. But I am also aware that I have been hearing this kind of thing all my life — about comic books, movies, television, and now multimedia. My children are all quite different from me and from each other, so comparisons are difficult, but I think that their mental faculties are no less acute than mine. They seem to have survived and triumphed over television. They may, while growing into adulthood, have spent less time reading than I did, but they also have learned more than I did at their ages. (I'm not being chauvinistic here. These comments apply to my children's friends as well.) I am content with the bargain. I am not afraid of the transition, especially since I will not live to see it completed.

Chapter 9

Modern Telecommunications: The Information Highway

I mentioned earlier that we are caught up in a rate of technological change never before experienced in the world. Telecommunications may be the technology that changes the fastest of all and with the greatest impact on society. In spite of the growth in the number of computers in use, we are not all computer users, but you virtually have to be a homeless person to live in North America and not be a user of telecommunications, whether by telephone, radio, or television. And those who use it for their business or profession, for social interaction, or for entertainment may find that telecommunication changes affect them more than any of the other technologies.

In my parents' day, a long-distance telephone call among siblings was almost invariably limited to bad news and it would not surprise me if none of my grandparents had ever made a long-distance telephone call. All of my and my wife's brothers and sisters live in different states or provinces and none of us live in or near our city of birth. The principal method of communication among us is the telephone and, very recently, fax. My four children all have access to electronic mail, which has become the dominant mode of communication among us.

In this chapter, we will briefly review how the principal communications technologies work, at a not very technical level. More information about telecommunications in general can be found in references cited in the notes.[1]

Some Telecommunication Basics

Telecommunication literally means "distant communication," but has come to mean communication by electrical or electronic means. Such other forms as carrier pigeon, smoke signals, or semaphore flags are means of long-distance communication, but have little significance in today's world. The two primary types of channels of telecommunication are wire, using electrical voltage as an analog of a symbol, and electromagnetic waves, which may travel as laser light on a fiber-optic (glass) cable or as radio.

The Electromagnetic Spectrum

The telephone system is the primary communication user of wires and cables, although of course the cable television industry, a relative newcomer, is catching up. Electromagnetic, or radio, waves are used by all modern telecommunications media to some extent. The electromagnetic spectrum is the range of lengths of these waves. They can be manifested as many phenomena, such as light, X-rays, or broadcast radio. Figure 7 shows some of the kinds of waves occurring in various parts of the spectrum, with their wavelengths and frequencies. Visible light uses waves of length from about 4 to 7 ten-millionths of a meter, or a frequency of about 430 to 750 billion cycles per second, or gigahertz (Ghz). These numbers are almost impossible to visualize. Infrared light (uses include transmission from your TV remote to the receiver or cable converter) occurs in the range from about 1,000 to 100,000 Ghz. AM radio uses 1 to 10 million cycles per second, or megahertz (MHz). FM radio and broadcast television use 0.9 to 100 MHz. Microwaves, which have a variety of uses, occupy the range from 1 to 100 Ghz. What is important to realize is that these different forms of electromagnetic waves have very different characteristics, but are all the same basic phenomenon and are used extensively in communication.

Fax is not a form of wave. It is a means of converting messages from one form (images on a page) to another (electrical pulses) and then back

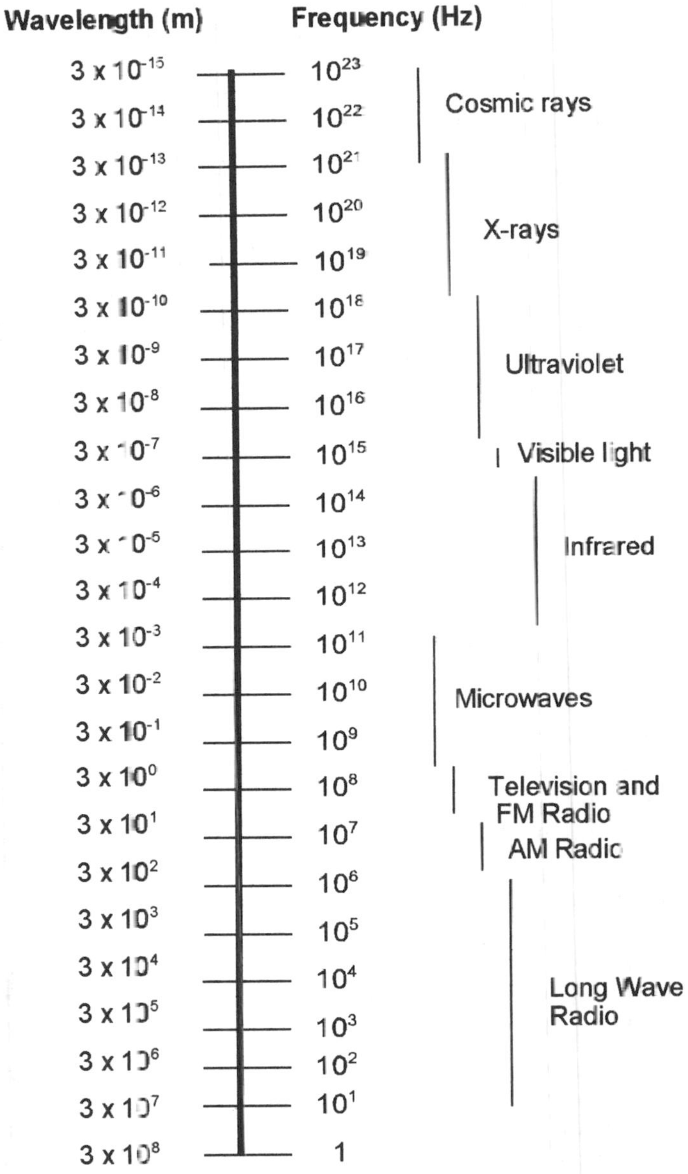

Wavelength (m)   Frequency (Hz)

Wavelength (m)	Frequency (Hz)	Band
3×10^{-15}	10^{23}	
3×10^{-14}	10^{22}	Cosmic rays
3×10^{-13}	10^{21}	
3×10^{-12}	10^{20}	
3×10^{-11}	10^{19}	X-rays
3×10^{-10}	10^{18}	
3×10^{-9}	10^{17}	Ultraviolet
3×10^{-8}	10^{16}	
3×10^{-7}	10^{15}	Visible light
3×10^{-6}	10^{14}	
3×10^{-5}	10^{13}	Infrared
3×10^{-4}	10^{12}	
3×10^{-3}	10^{11}	
3×10^{-2}	10^{10}	Microwaves
3×10^{-1}	10^{9}	
3×10^{0}	10^{8}	Television and FM Radio
3×10^{1}	10^{7}	AM Radio
3×10^{2}	10^{6}	
3×10^{3}	10^{5}	
3×10^{4}	10^{4}	Long Wave Radio
3×10^{5}	10^{3}	
3×10^{6}	10^{2}	
3×10^{7}	10^{1}	
3×10^{8}	1	

Figure 7. The electromagnetic spectrum. *The principal types of waves are shown, with their frequencies and wavelengths.*

again. Electromagnetic waves carry the information about the light level of each pixel, perhaps over microwave. The transmission goes through the telephone system. Satellites are a means of relaying microwaves, not a basically different medium of transmission.

In both these cases the original message has to be converted into a form capable of being transmitted over wires or cables or radio. We may talk about sending radio, TV, or microwave signals "over the airwaves." *Airwaves* is actually an inaccurate term. Electromagnetic waves can travel through empty space. They do not use air to make the waves. Neither, by the way, do they go "over the ether." This meaning of *ether* was applied in the 19th century to a medium assumed to be necessary to carry electromagnetic waves. Other kinds of waves require a tangible medium in which to travel: water, air, or the wooden sounding board of a piano. Hence, a material called ether was postulated, but turned out not to exist. Electromagnetic waves do not require a medium through which to travel. But we still sometimes refer to the airwaves or ether, informally.

At the receiving end of a transmission, the signal is reconverted to the form needed by the human or mechanical destination. Many modern systems use a variety of transmission forms before the signal gets to its final destination. Telephone uses copper wire, glass or fiber-optic cable, and microwaves, a form of radio. Television frequently goes from its originating station up to a satellite, then down to a receiving station that retransmits it to the home. That last link is likely to be by coaxial cable, an improvement on basic wire lines that can carry more information with less noise. More and more homes now have their own antennae with which to pick up the satellite's signals directly.

Bandwidth

In all telecommunication we are concerned with a concept called *bandwidth*. This is a measure of signal carrying capacity. By analogy, if we were dealing with the flow of water, the best measure of the capacity of a system of pipes to carry water is the diameter of the pipe. That does not measure the amount of water flowing at any given time, it measures the maximum capacity for flow. The TV cable going into your house has a far greater capacity for signal carrying than does the copper wire carrying your telephone messages. As routine communication into a household or office has escalated from voice telephone to telephone-carried computer messages, to television, to the Internet carrying everything, the demand for bandwidth keeps growing, just as the demand for greater capacity in water mains and sewers grows as a small community grows into a large one.

Why do we need so much bandwidth? A typical printed page of a 6-inch-by-9-inch book contains about 350 words and a typical word requires six to ten letters or bytes, including the space at the end, which must be explicitly represented as a character in computer memory. Calling it ten bytes for simplicity, we have $350 \times 10 = 3,500$ bytes per page. (A byte consists of eight smaller elements called *bits*.) If we were to take a picture of the page, in black and white, and using 300 pixels per inch in both directions (left to right, top to bottom), the page would require $6 \times 300 \times 9 \times 300 = 4,860,000$ pixels to represent it. The number of bits required for a pixel depends on the shading resolution required, but at minimum there must be at least one bit (indicating all black or all white), and there could easily be eight or more, to represent shading. Thus, the 4.86 mega-pixels could require that number of bytes. If the image were in color (the transmission system doesn't know this is a black-and-white page — it is set up for whatever colors we send), we need three or four times this number of pixels, for the three primary colors and sometimes black. Modern computers allow for *millions* of color and tone variations. If this image is to be transmitted as part of a motion picture, each frame persists for only 1/30 of a second. Thus, we have to be able to send 4.86 x 3 colors = 13.58 megabytes every 1/30 second, or 407 mbytes per second. Quite a load. It can be reduced considerably by compacting the message, but compare it with the transmission requirement for the digital text version of the page, 3,500 bytes. We keep moving in the direction of wanting more sound, graphics, and motion pictures in anything we read. This increase in demand seems incessant.

We can provide more bandwidth by increasing the number of lines in use or by increasing the signal-carrying capacity of lines. There are also ways of encoding or compacting the messages, which reduce the number of bytes needed, hence demand less bandwidth. One method, for example, sends only that part of an image that differs from the previous one. If compaction were used with page images, which are essentially black and white, not gray, a large percent of the information is simply a repeat of the value of the previous pixel. With clever coding the number of bytes required can be very much less than the number derived above.

If your family telephone is in continual use, between the teenagers and the computer, the usual response is to get one or more additional

lines. Simple enough, and relatively cheap. To bring in broadcast television, though, in current technology, we need a much higher bandwidth
than phone lines give us, so we have the higher-capacity coaxial cables
bringing in the pictures. Household phone lines are not adequate for really heavy-duty Internet uses, so offices may get "wired" with coaxial or
fiber-optic cable or microwave links, and the cable operators are casting
covetous eyes on Internet traffic. Finally, the telephone world, not wish-
ing to be pushed aside, is working on ways to compact messages so telephone, too, can carry video.

The Telephone

The telephone was originally conceived as a point-to-point means of
communication. One person would call another using a wire strung
between their two houses or places of business. For a brief period the
telephone was tested as a medium of mass communication, meaning that
a single originator transmitted simultaneously to a large number of
receivers. But it quickly reverted to point-to-point where it has essentially
remained ever since. Yes, there are conference calls but they constitute a
tiny fraction of the total traffic load. Today, the worldwide telephone
system surely ranks as one of our most magnificent technological
achievements.

Recent Changes

The major changes we have seen in this technology in its approximately
120-year history are: the spreading of the network to the point where it
covers just about the entire world; an increase in fidelity of the sound
carried (or absence of noise); the lowering of the cost (enormously lower
for long distance, but even some lowering, in constant dollars, for local
service); and the gradual untethering of telephone users from wires. The
increase in fidelity comes both from widening the bandwidth and
eliminating or reducing noise.
   Cellular phones, if we all had one, would free us from the agonizing
search for a public telephone when out about town or on the highway,
and let us keep in touch and be kept in touch with, virtually wherever we
go. I, myself, do not particularly treasure this form of progress, but I do
not work in a profession where I have to be in frequent touch with an

office. I do admit to some nervousness when the teenage drivers in the family had the car for a weekend out of town. But I don't want to be kept up with when I, myself, am away for the weekend. I also do not appreciate the sight of an auto driver with one hand on the steering wheel, the other on the telephone, and the mind on — what? Like it or not, though, this is a trend with the feeling of inevitability about it. We seem headed under full steam for a world in which we are always available by phone.

The telephone has also changed in terms of what can be connected to it. In the 1960s, AT&T in the United States, and its local subsidiaries, did not permit any device to be connected to their telephones without their permission. The world was just awakening to the joys of connecting a computer to the telephone. The issue was forced by Thomas Carter, the head of a company producing the Carterfone,[2] a device to connect radio with telephones. Carter brought suit against AT&T and, after a lengthy battle, won the right of relatively unhampered connection of non-Bell equipment to Bell telephones in users' hands. Things just took off from there. Today, we have telephones made by many different companies connected to AT&T lines and we have competitors for the lines, too. Computer and fax users and manufacturers are probably the great beneficiaries of this. Today, the telephone network is used in ways that Mr. Bell, and maybe even Mr. Carter, could never have conceived.

Telephone Mechanics

When we use the telephone, the input sound is converted to electric current inside the telephone mouthpiece. This current, which may be amplified along the way, goes eventually to the receiver of the destination's telephone, where it is converted back into sound. So, we do not literally send sound over all those telephone wires. Neither do we send radio waves, in basic telephony. What travels is an electric current analog of the sound.

Somewhere in your part of town there is an innocuous building identified as belonging to your local telephone company, but having a rather forbidding look. The public is not invited in. This will house a switching center or exchange. Your telephone is connected to it by wire actually a pair of wires called a *twisted pair*. We can avoid all sorts of technical detail if you will accept that there are two wires to help reduce the effect of externally generated noise. When I occasionally use one of the two phone lines in our house to call someone upstairs on the other line, the call goes out of the house to the exchange, then back again to the third floor of my house.

If I call someone in another exchange, my call goes to my exchange, to the other exchange, and then to the callee's home or office. If it is a long-distance call, it may be routed through a large number of exchanges. This all happens quickly and without the caller being aware of the activity. In the old days, each segment of each call would travel on a copper wire. In the really old days, a human operator at each switching station had to make the connection from an incoming line to an outgoing one. Now, it is all automatic.

Today, it is possible to *multiplex* calls, so any cable can carry more than one call at a time, and part of the transmission may go part of its way via fiber-optic cable or microwave, both having greater capacity than the old wires. The multiplexing method is somewhat like what we use to enable many radio or TV stations to broadcast in one city, without inter-fering. Each is assigned a base frequency and it must broadcast on (actu-ally near) that frequency. If everyone does it right, there is no interfer-ence among all the simultaneous broadcasts. Similarly, each of several phone messages can be transmitted over multiplexed lines on a different frequency, thus economizing on lines, yet minimizing interference among the calls.

If all telephone links were of copper wire strung on poles, we would hardly see the sky in our larger cities. Some lines go underground and some undersea. The first trans-Atlantic telegraph cable was laid in 1866, but the first telephone cable across the Atlantic did not come until 1956. Prior to this, trans-Atlantic calls were possible but went by radio, which was an expensive way to do it, because time on the transmission facilities had to be scheduled in advance, which often delayed the start of a call. Hence, it is not surprising that there was not much over-the-ocean calling back then.

More and more, except for that first link connecting your home to the exchange, telephone messages travel by radio, using the microwave por-tion of the spectrum. You have probably seen clusters of strange-looking parabolic or oval shapes on tall buildings or towers around town. These are microwave antennas. They are much cheaper than are thousands of telephone poles and tons of copper. The glass, or fiber-optic, cables offer a very wide bandwidth compared with copper and receive much less in-terference from external noise.

Telephone via Radio

To make use of a microwave link, whether terrestrial or extraterrestrial, the electric current analog of the sound is converted to electromagnetic radio waves by changing the frequency or amplitude of a basic wave according to the amplitude and frequency of the sound. This is *modulation*. The receiver *demodulates* the radio waves back into electric current and the speaker or earpiece of the telephone converts that back into sound. While in electromagnetic wave form the signals can leap tall buildings, mountains, and rivers, which poles carrying wire cannot.

Locally, we now have *cells* that contain radio receivers and transmitters. When you initiate a cellular telephone call, it is transmitted by radio to the cell nearest you, then relayed to the cell nearest your destination. The cellular system tracks the movement of the caller's phone by noting which cells can hear it best, and it hands off the call to the cell into which the caller is moving. This is done without the caller being aware. A portable telephone is a sort of primitive version. It has a base set that does not move as you walk around house or yard, but you must stay within some specified distance of your base set. Transmission from you to the base is by radio; from the base to the telephone exchange is by wire. The point of all this is that radio and telephone have come together in many ways and the consequence is loosening the bonds of the telephone wire on our lives. A more modern version of the cell phone is the Personal Communication System (PCS) which is similar to cellular but uses a higher frequency range, giving more bandwidth and allows for combinations of voice transmissions with fax and computer e-mail connections.[3] There is no fundamental difference between cellular and PCS. In fact, service providers are tending to refer to both as *wireless telephone*. There are some high-frequency allocations for PCS but the lower frequencies may still be in use.

Radio and Television

Radio went from a primarily point-to-point system, or one used for broadcasting to a small audience, to a largely mass medium. We mentioned in Chapter 3 that the ship *Titanic* broadcast its desperate calls for help by radio, but it could not have expected to benefit from more than a few nearby ships and, since there were no air rescue services in its day, it hardly mattered if distant receivers heard the messages.

Entertainment radio attracted large audiences for its day. People in the 1930s and 1940s stayed home to hear their favorite programs. Families listened together. There were some social changes as a result of radio. It was the first time live entertainment and news could consistently be brought to a group in the home. Families began to gather together to hear their favorite programs. Political campaigns began to cater to candidates' media images. The social change was not as intense as that later attributed to television, which became a serious commercial product in the late 1940s and early 1950s. At the end of World War II, production facilities were quickly retooled for consumer, rather than military, production. People had accumulated money during the war when there were few consumer goods to buy. Television sets sold like hotcakes. Ever since, TV has attracted not only large numbers of viewers but large numbers of studies and criticisms. Whether it is the cause of our more violent society, whether it has lowered the attention span of people raised on it rather than on reading, whether it has raised our expectations for money and success to unattainable levels, whether it has destroyed interest in reading, it has been accused of all these sins. It does seem to have had an enormous impact even if we are not always sure what that impact has been.

Radio and TV Mechanics

Radio works much like telephone with microwave links. Sound becomes electric current in the microphone and shortly thereafter becomes electromagnetic waves in a transmitter. The antenna of the home receiver converts the waves back into current and the amplifier and loudspeaker finish the job, reconstructing sound. With all these conversions and reconversions and trips through the "airwaves," it is amazing that the resulting sound can be so true to the original.

In television, the sound is sent separately from the image, traveling as FM radio. The image is scanned, as described in Chapter 5. If we stay with black-and-white TV, to keep it simple, the amount of light sensed at any point in the image is converted to an analogous electric current, which is then, as with sound, converted into electromagnetic waves for transmission. The receiver, again, reverses the modulation of both sound and image. The sound comes out of a loudspeaker and the image signal goes to an electron tube where the intensity of light in the original governs the intensity of a spot on the inside of the face of the tube.

Cable TV

A master antenna can pick up broadcast television signals, demodulate them, and send them out via cable to multiple TV receivers as electric current. In a sense, the only way this differs from each of us having our own antenna is that the cable operators have a bigger, better one. Where our antenna's access to a signal might be partially blocked by a large building, theirs might be on top of the building. The cable is a very high bandwidth channel that can carry a large number of converted signals simultaneously. So, we get better-quality images, and we get more of them than most of us can pick up directly.

As one community after another becomes wired for TV cable, there is created a new communications network, capable of rivaling the telephone net in terms of its near-universal reach, but of much higher signal-carrying capacity. This new net could carry business and other transactions as well as entertainment — shopping, remote banking, video on demand, and the like. Cable and telephone operators will compete vigorously for the right to carry these services. One such service could be the delivery of the equivalent of our newspapers, magazines, or books. They could be recorded on paper at home or read in audio or video form.

Satellite Communication

In 1945, Arthur C. Clarke, an engineer who was to become primarily noted as a science fiction writer, got the idea to put an antenna in an artificial satellite in space, and this led to our present communication satellites.[4] It accelerated the spread of the telephone network, and its interconnected computers, around the world.

Modern communication satellites are a combination of an antenna and a relay station. The satellite is in what is called a *geosynchronous orbit* — that is, it is so synchronized with the earth's rotation that it appears stationary to us, observing from Earth. It does not rise and set. It always seems to us to be in the same place. (See Figure 8.) Typically, these satellites sit about 35,000 km, or 22,000 miles, above the earth's surface. They receive microwave signals, amplify them, and retransmit them back toward Earth.

Today, a satellite can handle the equivalent of about 15,000 telephone circuits. Because a call to the house next door would have to go up 36,000 km and then down again, the horizontal distance from my house to your

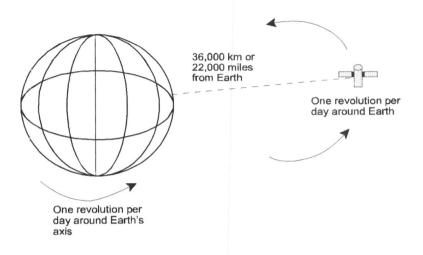

Figure 8. A communications satellite. *The "bird" is in orbit about 22,000 miles above Earth and appears fixed above one spot on the surface.*

house makes very little difference in computing the total distance the messages must travel. Whether we are 1 km or 5,000 km apart, the up and down distances dominate. Hence the cost of transmission is generally independent of the ground distance between communicants. Think what the equivalent capacity would mean in terms of poles and wires. Think also what it means to telephoning that distance ceases to be a real factor in cost. These figures, by the way, change often, always in the direction of more capacity, more speed, and lower cost per message.

Satellites can also relay radio or television signals, since these, too, are microwaves. Before we had satellites, TV signals going from a studio in New York to a station in San Francisco for rebroadcast to the latter's local audience required a large number of phone lines to convey the complete signal, each line having a limited bandwidth. Similarly, broadcasting, say, a live football game meant a large number of phone lines from the stadium to the studio for transmission to the network. Today, the TV company brings a truck-mounted antenna into or next to the stadium and sends its signals directly up, then down to whomever has contracted for it.

Satellite TV has been called the "death star" because it poses an economic threat to the continued existence of cable companies. Using a small receiving antenna in the home, it can bring in a large number of channels, far more than cable is now offering. The antenna costs more than a cable converter (now built into most TV receivers), but operational costs are about the same.

The effect of satellite TV on print publishing is not likely to be so direct or dramatic. In fact, the satellites cannot do anything the cables cannot do other than offer more channels. Perhaps their impact on print might be that they increase our dependence on TV for news, entertainment, and even education.

Facsimile

As we said in Chapter 5, fax is a communication mode that, like part of television, converts an image to an electrical analog. Then fax sends this analogous signal over the telephone system, rather than directly by radio, to the receiving fax. If the receiving machine is actually a computer acting like a fax, it might contain software capable of recognizing the form of individual characters of a message and converting them into digital form. This, recall, is optical character recognition. Lacking such computer programs, as most of us do, our incoming fax messages, even if received by the computer, are stored as images — pictures of letters, not codes for letters.

The fax machine is a relatively simple one as modern electronic devices go, but it has played a big role in changing the way offices are run. Although a fax message requires a phone call, often a long-distance one, it can be cheaper than conventional postage because it need not be placed in an envelope, addressed, stamped, and taken to a post office. Its speed, compared with post, is often worth the price, even if it were to cost more than postage. As these machines become more widespread, the tendency to use them in new ways increases.

An interesting social change is that fax messages containing an image of a signature are now often accepted as fully legal, where previously only an original signature was allowed. In effect, a picture of the signature is accepted. Since we now have the ability to scan images, store them in a computer, and copy onto paper or send as an outgoing signal, it means that we can easily copy an image of a signature from one document into another. It is surprising, at least to me, that this has gained such widespread acceptance. Apparently, the convenience and speed fax gives us seem to have overcome our traditional insistence on the written word, and the literal truth of the written original. Exactly this issue has been raised (see Chapter 4) as an objection to electronic documents — that the concept of the legal original may cease to exist.

The Internet

While the telephone network was the telecommunications phenomenon of the late 19th and first half of the 20th century, followed by radio and television in the second half of this century, the Internet[5] is clearly the "phenom" of the 1990s and possibly beyond. Computer networks as we know them today date from the late 1960s. Gradually there got to be more and more of them and, like national telephone companies, they began to interconnect. For dissimilar systems to interconnect there is a need for standards. For example, in North America, electrical outlets in homes and offices are all more or less alike. Overwhelmingly, they are for 110-115 volt alternating current or 220 v for heavy-duty use. The connectors or plugs that link the 110 v outlets with the appliances, while varying somewhat, all accept the old-fashioned, two-pronged connectors or the modern three-pronged ones. By contrast, electric light bulbs or lamps, having once achieved a high degree of standardization, seem now at the mercy of the makers of designer lamps and they come in all sorts of sizes and shapes. These differences can drive a householder to distraction when trying to find a compatible replacement. Standards for interconnecting components eliminate such problems. They are just as necessary for electronic systems that exchange messages with each other.

As computer networks grew in size and number, it became apparent that there had to be a standard way of addressing a message and a standard way to organize the information in a message, so every network could read incoming messages from another net. Note that this does not refer to content. There are only standards for organizing a message, or for how an address is represented, not for what you say. An ordinary social letter will have a return address, a date, a salutation, the body of the letter, a stereotyped terminating statement such as "sincerely yours," and a signature. This form is more or less "required" by social custom, but the custom does not dictate content. There are other customs and some laws that do relate to content and these are often hotly contested today.

The Internet is not a giant network with a central management. It is a set of interconnected smaller networks all following standards agreed upon by the individual networks. There has to be an organization to create the standards but there is no central management or ownership. That, in itself, is a remarkable feature, that such an organism has grown up and become worldwide and highly successful without central or hierarchical management.

The Internet, then, is a *means of communication.* Its communication channels are likely to be provided by a telephone company, although cable TV operators are also trying to get into the act. One of Canada's largest cable companies began Internet service provision in 1996. It is important to bear in mind that the Internet is *not* a producer or storer of information. It is not technically correct to say that such and such information is stored on the Internet any more than good, juicy gossip is stored on the telephone. Messages are stored in computers, which are linked into one of the networks of the Internet.

"Surfing the net" means browsing through the nodes, which are computers connected to the smaller networks, which, in turn, are part of the Internet linkage. The computers must have software that does the searching or manages the storing of information. Not that it is terrible to make a slight misuse of words and call the computer the Internet, but it can create an incorrect image and can lead to the false impression that having an Internet or having the means of access to it are all that are needed to have access to the information. It also requires that the people who control the nodes be willing to make the information they have available. And when they do this, it is not all going to be free, or not free forever. There is also the requirement that people who want information know how to find it. As the web continues to add more records, users will have to learn to be more precise in asking questions or be swamped by what they retrieve. This means the web browsers will have to provide search systems that allow for this greater precision and users will have to learn how to use them.

Who is responsible for providing the information and the necessary software? It is *not* the Internet organization. It is the individual members of individual member networks. It costs money, time, and effort to provide this information. It costs money, time, and effort to search for it.

Internet Growth

How large is the Internet and how fast is it growing? Such figures tend to be obsolete the day after they are published, but Figure A16 in the Appendix shows some of its phenomenal growth in terms of number of domains. A domain is an addressable set of files and programs on a computer, and there can be more than one domain on a single computer. The domain may, in turn, deal with many users. Be cautious in interpreting these figures. It looks like we are headed for an infinite number of users, but there has to be a limit. However rapidly that number may be growing today, it must inevitably reach a saturation point. There are only a finite number of people in the world. Telephones tend to be associated with a home or office. At home, we tend to have one telephone number per household, occasionally one or two more, especially when there are highly socialized children. In offices, we are more likely to have one phone line per person. These are *lines*. A modern household may well have several *instruments* connected to any one line. Internet access is likely to be assigned to a person, rather than a household. You become an Internet user by getting an individual password, not by ownership of a computer. You can use the password in the public library in many communities, or members of a family may share a single computer among multiple password owners. But usage is limited absolutely by the number of people in a network's area of service. It is also limited by the number with computer access. And in spite of the growth in number of such people, that market is not yet saturated.

The World Wide Web

It started slowly, but by now a great many newspapers and magazines are available in one form or another through an Internet service called the World Wide Web (WWW), which is also more a standard than an organization. It is a means of searching and transmitting multimedia works with emphasis on those having a high degree of hypertext inter-connectivity. It was developed by Tim Berners-Lee,[6] a British physicist, as a means of exchanging information among scientists.

The web consists of a network of *sites* or *domains*, where a site is approximately a computer, but actually there can be more than one site per computer. Each site has an address and, just as with a home or office, the address tells how to find a family or organization but does not tell the names of all who live there. Within a site may be many electronic equivalents of documents. The user (web aficionados like to talk in terms of *visits* to sites, hence a user is a *visitor*) starts with a *home page* that normally contains a title, a brief description of what is to be found there, and a menu of specifics at this site. The visitor can select what is wanted for viewing. Any document or menu item can point to another document at another site. If, for example, I am reading an article in an electronic journal and it cites another article, in a different journal, a few clicks of my mouse and I am visiting that other journal.

At the present time, much of the material on the web is still experimental. Some people have a personal home page. Some publishers will have many documents available, or only a sample to lure subscribers to the printed forms. The commercial publications on the web today are often free. Will the publishers continue to provide their work for free? The more the world moves into electronic publishing, the less likely it is that these services will be free of direct user cost. How can they be? How can a for-profit publisher survive by giving away its product, except as a temporary loss leader? Or, perhaps like some trade publications, they could turn a profit solely from advertising. But that, too, is a price the consumer pays in terms of time and, indirectly, by buying advertised products and services that advertising brings to our attention. Lengthy phone calls to the Internet *do* cost more money than old-fashioned voice calls. One way or the other, we will end up paying for them.

Today many of us are fascinated by our ability to get on the web and search for documents (or pictures or sound cuts) by, about, or from some person, place, idea, or event. One of the great fascinations, in fact the driving force behind creation of the web, is that if one document cites another, then wherever in the world the cited document may be stored, so long as it is also on the web, we can get to it in a few seconds. For research, whether in science, law, or the making of spoon bread, that is really quite a service. Once you've experienced it, you will not easily give it up. The creation of a habit-forming service promises good things for those who sell it.

The Internet and its World Wide Web have changed the communication habits of many people, but have not yet really changed publishing, for three reasons. One is that users of information have the habit of getting it from the web at no charge. Second is that the means of exchanging money — paying for goods or services — is not yet well established. The third is the actual transference of the goods — books, magazines, motion pictures, etc. We cannot yet end up with as nicely packaged a result as a well-printed and bound book or magazine.

Since so many publishers have made information available free during what, to them, is an experimental period, they may find it hard to start charging for it as the electronic means of distribution catches on.

We have precedent for "exchanging" money by providing a credit or debit card number, which would be verified before the transaction were fully approved. Recent efforts by Visa, Master Card, and Microsoft have led to a secure means of transferring most credit card information. This will probably make network money exchanges more popular. Another approach is to require that the buyer have established an account directly with the seller and use that account number, with billing and payment done in the conventional manner. This is already done with other on-line computer services (CompuServe, Dialog, Dow Jones News Retrieval, etc.) and is essentially the basis on which we deal with our local telephone company.

We do have some electronic journals today. These are periodicals that do not appear in print but only in electronic form. They work well for fast dissemination, but they generally lack the prestige of established print publications. Prestigious authors have little incentive to publish in unknown journals. More on this in Chapter 10. Transferring, or "downloading" recorded works such as books or articles can be done now, but not in packages as neat as today's books and magazines. Most computer printers are dedicated to 8 ½- by-11- inch paper or its European near-equivalent, the A4. Computer paper is not bad for printing, but it is not the slick, coated material on which color photographs appear so vivid, that is used in publishing. And for binding, there are inexpensive binding devices, beginning with the stapler, but they hardly match the quality and style of traditional bookbinding.

The net is bound to become more commercial, in time. When it does so, its "publications" will be generally accepted and the mechanics of use will be either simple or learned by most people at an early age. It will happen, but we have not worked out all the kinks yet.

Every day seems to bring more new material to the web, making it increasingly attractive. The more attractive it is, the more customers are drawn, and the more customers, the more content that will be offered and the higher the price that will be charged. Pricing in telecommunications is in a sense relative. Technological developments tend to make transmission less costly per byte so communications carriers can usually make more by holding prices constant and lowering the cost of service. But such developments also tend to encourage us to want to transmit more.

Multiple Channels, Multiple Modes

Telephone for a century has been a medium for voice communication. Radio started life as a point-to-point medium but its dominant use today is in broadcasting. Television has been a mass broadcast medium for its whole existence. (A use such as closed-circuit transmission for building security is a very minor part of the industry.) There have been attempts at two-way or interactive broadcast television, but they never caught on, probably because they involved the same kind of programming in a different media environment. There have been experiments with picture phone, telephone service in which the parties see TV images of one another while talking. This never caught on, either, but is being revived coupled with various forms of e-mail. Groups of people working together through the medium of a computer are becoming popular. This involves exchanging text, graphics, and video images.[7]

But now we are on the verge of a massive coming together of media. Fax uses the telephone system and by now probably most fax machines have a voice telephone attached and some have a switch to direct incoming calls between the phone and fax. Now, voice conversations about a contract can be immediately followed by a signed copy of the contract. If we were to use the TV cable for two-way communication, there is enough transmission capacity that we could easily see the person we were talking to *and* send copies of documents. We can feed television images into a computer and transmit them remotely. While no such complete public multimedia telecommunications system yet exists, it is technologically feasible. While there is no existing demand for it, it might create its own demand as it is brought into existence.

What if multimedia becomes the standard for interpersonal telecommunication? If it does, would that not indicate a probable demand for multimedia in other forms of communication — education, entertainment, enlightenment? These are the prime applications of print media. I cannot but think it must, or anyway that it will. That means that this drive to new or combined forms of routine telecommunication is going, some day, to drag the print media along with them.

The Information Superhighway

I do not know exactly what this term *information highway,* or *information superhighway,* means and I suspect no one else knows, either. It seems to have been coined by U.S. Vice President Al Gore in 1990.[8] I find it half the time used to refer to the Internet[9] and the other half to the existing and potential commercial telecommunications facilities such as communication satellites and cable TV. Maybe it is not so surprising that there is this confusion. Whatever it is, in one sense the superhighway does not yet fully exist or, in the case of the Internet, it exists in a frequently changing state not yet universally available.

In another sense, the information highway can be said to be all the collective means a society has of transferring information, and in this sense it has always existed, just as we have "always" had roadways, if we accept everything from a footpath through the woods to an interstate highway as a road. We have always improved our means of transportation — of people and goods and messages. The modern developments of the Internet continues that tradition.

I think it is significant that even the notion of an information superhighway gets so much press and other media attention. There seems to be a feeling in the air of a great social change coming. This may become a self-fulfilling prophecy — there will be a great change because so many people believe it is coming and act accordingly. Perhaps, on the other hand, we have a genuine technology–culture movement taking place because the world genuinely wants it. There can be little doubt that there are vast sums to be made from this change by telecommunication, computer, and related companies and that perceived value to us, the users, has to be some part of the reason for our headlong dash toward the brave new superhighway. It is not inconceivable that all of what we now think of as publications will come to us along a future electronic information highway. It is also not inconceivable that one day we may become so sophisticated in understanding information that we see this highway as almost a natural thing, and include in it all the means of distribution of books, newspapers, notes, or letters.

Chapter 10
Distribution

Distribution is not the same as telecommunication. Telecommunication is a means to an end and that end is distribution, delivering the right messages to the right destinations. The purpose of publishing is to get information into the hands of users. The technological means of doing so are important, but are not the only considerations. In the days before movable type, the means of distribution were as limited as were the books available to distribute. Victor Hugo's *Notre Dame* contains several references to the printed book as replacement for the cathedral as a source of information for the people, and Napoleon Bonaparte is quoted as having said, "Cannon killed the feudal system, ink will kill the modern social organization."[1] Today, we have the dazzling options described in the previous chapter, which may yet lead to another great social disruption. But just as important as the means of communication are time, cost, selectivity of distribution, and selectivity of access by users. Let us consider these factors separately.

Time

News travels very fast in our time. We are a long way from the War of 1812 when the Battle of New Orleans was fought after the war had officially ended, the news not having reached the combatants in time.[2] When the news of Wellington's victory at Waterloo first came to England via Nathan Rothschild's private courier service, and thence to the government, it came so fast that it was not believed until slower confirmation came along.[3] Think of the advantage to the merchant or stock trader who alone knows a major war-ending battle has been won. Yet, the importance of rapid communication seemed not to be realized by the British government thinking in those days. Compare with the media coverage of the Gulf War in 1991, when CNN broadcast the war live to both sides. Probably today, the immediacy of news is one of the great drawing cards for TV newscasts. When it comes to interpretation or historical perspective on news events, we are usually content to wait until books are published, months, years, or even centuries after the events.

The sense of urgency, then, is not constant. It depends on context. It does an investor little good to know the price of a share of stock a year ago, when it would normally have been a routine news item. But we are still mulling over the origin of Stonehenge and the character of Christopher Columbus. There is no rush to settle these issues, as there is to know who won this afternoon's ball game.

In recent years there has unquestionably been an increasing demand for speed in most business or administrative activities. Instead of automation gaining us more leisure, it has tended to gain us more pressure to do everything faster. We were once content to borrow books or journals from a distant library via slow-moving interlibrary loan if these texts were not available locally. No longer. Now, we tend to want a photocopy tomorrow, or a web copy this minute.

For those in a hurry, the Internet and its World Wide Web is just the ticket. And every prediction of the future development of telecommunications promises us wider bandwidth at relatively small extra cost. But sometimes the urgency does cost more or it gives us information in a form other than what we really want. Publishing in a hurry reduces time to evaluate and verify content.

Cost

We are on the subject of the cost of distribution, not of production. And here, electronic publication has an immense edge on the paper form. Sending the text of a typical newspaper or magazine article over the Internet costs per nies. Getting that article into a reader's hands in print form usually involves a larger package (the entire magazine), postage, packaging, and handling by the shipper, and handling by the recipient. Libraries receive a great many periodicals every year and just checking them in and making claims for missing copies is an expensive process. This is all exclusive of the cost of delay of receipt of content, if any.

Downloading a complete book or an article with many graphics included can take more significant amounts of time and demands higher cost equipment of both sender and receiver. But such large messages can be compacted. If full-length movies can be sent economically over telephone lines, so can books. Compacting, in this sense, does not mean condensing à la the *Reader's Digest.* It means encoding the text or pictures in such a way as to reduce the volume of symbols to be transmitted, but enabling the original to be fully reconstructed at the destination.

Putting a document in a form suitable for transmission in the first place costs something. It is ever less because printing of books and journals today almost invariably means that the text has been put in computer-readable form at some time. As noted in the discussion of multimedia, the cost of producing a multimedia "book" will be more than that for a conventionally printed book, but the cost of distribution can be much lower.

It will be possible, again as we discussed in Chapter 9, to distribute many publications directly to the user, bypassing newsstands and bookstores. While this is not yet practicable, it could become so. Many of us would regret the loss of the bookstores as a collective institution. They might save themselves by offering more services and relying less on pure sales of packages of paper. More of that later.

We might give a thought to the distribution of costs, in terms of division of charges among users. When an instructor selects a book to be used as a text in a university course, the student is normally expected to buy the book. A few can get by with the library's copies, but in the United States, libraries normally do not have enough copies of any book to satisfy a whole class. So, the student pays. In Europe, students do not generally buy books to read for a class. Hence, the number of sales is lower and the price higher than in the U.S.

Recently, instructors have taken collections of articles from journals or chapters from books and compiled them into a customized work, tailored to a specific course. The student pays again, but the costs are not high because the printing is relatively rough (photocopy vs. printing press), there is virtually no distribution cost, and the editorial work has already been done. Lately, instructors have taken to putting class materials on computer networks.[4] With these, the student does not pay. It's free. Well, as the old saying goes, if you believe that, I have a nice bridge leading to Brooklyn that I'd like to sell you. It's not free at all, but the cost is bundled into tuition, grants, and the like, and so well spread among many payers that few notice it. It makes no economic sense to assume that all these services will continue to be offered forever at what have been come-on prices. Electronic distribution can remain cheaper than shipping large quantities of paper, but not by the infinite ratio of some finite cost to what now appears to be no cost at all.

In November 1995, Bell Canada announced it would offer Internet services to its customers directly, not requiring them to sign up with any new communication services beyond the phone company they already dealt with. The announced price for their service was quite a bit higher than the average charged by independent providers, and industry watchers wondered why. A few days later came a possible answer.[5] Bell announced its intent to raise the price it charges to the independent service providers. Then it announced that its actual charge for its own service would be about that currently charged by the others. This is not necessarily highway robbery and it is not necessarily unwarranted. As noted in Chapter 9, the cost of local telephone service is bound to go up as we lower long-distance rates and increase the number of long, computer-to-computer local calls. It is also not unreasonable to infer that these Bell Canada pricing changes have more to do with eliminating competition than with recovering costs of providing telephone service to independent Internet service providers. About that time, Rogers Cable, one of Canada's largest cable television operators, announced that they were going to offer their Internet connection service at around $40 per month, but with very high-speed service.

The idea that large volumes of data can be distributed free is doomed. There is no free lunch. Somebody pays.

Just to illustrate how much things have changed, in 1981 a British publication was poking fun at one of its government's developments, an on-line information system called Prestel.[6] This was an early attempt at the sort of services now provided by the Internet. It carried news and some commercial information such as airline schedules. It was updated daily, as I recall, but it did not have much detail. I remember a travel agent using it when I was planning a trip. He could find out about scheduled flights, but not the actual availability of seats. To do that would have required the database to be immediately updated after each reservation transaction, which Prestel could not do. So, the agent could tell me there was a scheduled flight from London to Geneva on such and such a day, but when I asked if there were seats available, he had to make a long-distance phone call to find out. The system was not well matched with users' needs. The published article compared Prestel with a newspaper, the *Financial Times*, and showed that the amount of data in Prestel equaled that in 80 issues of the paper. The cost at the time was overwhelmingly in favor of the *Financial Times*. And the once-a-day updating meant that the paper was not behind Prestel. The point was that at the then-current state of technology, user skill, and willingness to indulge, the paper could do a more cost-effective job of bringing information to users than the electronic system. The paper could have published schedules, too, but not seat data. Instant seat availability data, countrywide, is only possible with an interactive electronic system. Without it we risk either massive overbooking or massive under-loading of aircraft, each a high-cost outcome for someone.

Pushing and Pulling

Among the newer buzz words of the Internet aficionados are *push* and *pull technology*. Push technology is that which finds information for you and "pushes" it toward you. Pull technology is at your service, awaiting your specific request or command to pull information to you. These are quite useful concepts but they are not clearly distinct from each other and they are by no means new.

Pull technology is otherwise known as information retrieval. The user tells a search system (whether a person, card catalog, or computer system) what is wanted and the system finds it for him or her. We have pointed out that as illustrious a person as Bill Gates seems to feel that future computers will find what we want even when we do not know for sure ourselves. In fact, information retrieval is often a difficult process of negotiating a vaguely held sense of what is needed with what are sometimes difficult to understand statements of what is available. Pull technology allows you to be very specific because you are doing this search one time only, but it may be hard to say the right specific things given that you do not know about what is available.

Push technology, in my own personal experience, has been around for over 40 years. Basically, it means that a user creates a statement of general interest and leaves it to an agent of some sort to scan incoming information and deliver only what is wanted. Since you are describing the content of information that might arrive in the future, you have to be broad in your descriptions. In the early days this might have been as simple as a mail room sending incoming copies of certain journals to those who requested them. More elaborately, it can mean a panel of readers who read incoming mail to an organization and distribute copies to individuals whose interests, known or on file, match the documents. Today, you might leave such a profile with Pointcast or Dialog and you would get new information that meets your specifications.

Warning: it sounds great. It can be very helpful. It does not always work well. These interest profiles tend to be very general and they may not catch certain items that are new, that you did not think to specify an interest in, but are now important. This is not criticism. Such systems by their nature cannot be perfectly up to date. It is important to realize that push and pull work by having the intended recipient make a statement of interest. The system then either scans only newly arrived information (push) or everything in its collection (pull). The real differences are in how you state your interests and what set of data is searched.

Selectivity of Distribution

By selectivity of distribution on the part of the publisher, I do not mean any nefarious political scheming to limit who can read materials, but publishers' ability to target materials. Just as a simple example, supermarket book racks are unlikely to contain many of the Great Books, and the better university bookstores will, in fact, have more classics than they have Gothic romances. This is simply good marketing — a form of push technology. Put the books where there are most likely to be buyers. Frank Ogden's book *The Last Book You'll Ever Read and Other Lessons from the Future*[7] urges authors to self publish to avoid the publisher getting too big a share of the revenue. How the author who is not a professional book marketer is supposed to find the audience and inform them of the existence of the book is glossed over. It is interesting that his own book was put in the hands of a commercial publisher.

Journals and Magazines

Journals — scholarly journals, magazines, and newspapers — constitute a category that could gain significantly from electronic distribution. Few of us read any of these publications from cover to cover and we might be grateful if we could be offered the opportunity to pick up only those articles of interest from the Internet. Journals of all kinds tend to gain from illustrations, so they probably are better suited to the multimedia form than any other category but reference works. The fun of browsing might be lost, but actually it would only be made different, not lost.

Scientific and trade journals are a class of publication many of whose users look forward to the speed and convenience of electronic publishing. Production time for an article in a typical scholarly journal can take six months to more than a year from receipt of a manuscript from an author to its publication. Because articles in such journals are usually peer reviewed, it may take even longer. Peer review involves the editor sending a manuscript to several reviewers who are believed to have some special knowledge of the subject of the paper. The editor has little control over these reviewers. If they take six months (yes, what should be an evening's work can take that long), all the editor, author, and readers can do is wait and perhaps start over with a new reviewer. In some fields, preprints are sent out in something like journal form.[8] These are papers not yet refereed, but in certain subfields of physics, biology, or medicine, no one wants to wait. The disadvantage of this is that the slow process provides better quality control. The rush to publish or publicize brought us the "news" of the achievement of cold fusion,[9] the modern-day equivalent of perpetual motion as a goal of science or pseudo-science.

More and more journals are appearing in electronic form. While the leadership came from scholarly societies, some commercial publishers who publish large numbers of scientific journals are also experimenting.[10] In the early days of electronic databases, all we could get were bibliographic files, i.e., those that told what was published, perhaps even including an abstract. But the availability of full text, illustrations included, is very new. At this time, the economics are not settled — who will pay, how, and how much? It is so nice to get one's hands on an article in just a few minutes that this seems destined to be the way of the future. For all the high-speed delivery, though, readers still seem to like to read from paper. In an article describing some of these new journal distribution systems the head of Carnegie-Mellon University's library is quoted as saying, "We do not have strong evidence that people want to leave their hard copies. You put up a [local area network] and what do people do? They print out their e-mail."[11]

It is interesting that one of the major objections to electronic journals has to do with academic tenure.[12] Assistant professors are expected to publish (or perish) and the quality of the journals in which they publish is an important consideration. But the major journals are not stampeding into electronic publishing. The authors, then, have a dilemma — try for the most prestigious journal or the fastest route to publication? Critics of tenure committees' lack of reverence for electronic journals have a tendency to talk about transmission media, not content. A journal is neither good nor bad because it is distributed via the Internet. A journal that puts the content of its printed version on the web, after print distribution, is not speeding communication between author and reader. To do that it would have to make the content available before printing. Journals tend to be judged on past history of publication and, whether electronic or paper, it is hard to credit a journal that has no history. This is not to deny that we academics can be slow to adapt to new forms in life. The other side of this coin is that failure to announce new findings in a timely manner can be a disservice to science.[13]

Recently, McGraw-Hill's chairman expressed an interest in electronically publishing scholarly journals.[14] Reed Elsevier, a giant European journal publisher, is also interested in the transition,[15] as is the large book and journal publisher, Academic Press, and the professional society Association for Computing Machinery. Publishers are noting their customers' interest in the Internet and their authors' frustration with publishing delays. If the big names move into the field, then the hesitancy of authors to submit articles to no-name electronic journals might disappear because the major journals would be electronic, too.

What these ventures usually are, so far, is an attempt by a publisher to make its paper journals available to all subscribers in electronic form without loss of revenue. Typically, the user is required to have a subscription for the paper form. That reduces risk to the seller while giving the buyer the advantage of both forms of publication. For a library wanting to subscribe, it could mean less spent on binding journals and virtually no risk of theft or loss of copies.

Newspapers

I suspect the word *newspaper* will survive the physical artifact, just as we still "dial" a telephone number or "ship" a package by air. Technologically, newspapers would be the easiest of all the major forms of publication to convert to all-electronic form. The stories are short. There are graphics, but they pose no great challenge to modern technology. Subscription could be by subject, possibly covering several "papers," or it could be to the conventional packaging of a variety of subject matters in each daily edition. An advantage to this method of getting the news would be that we could easily call up old material to get background on a story, or switch to other databases, such as encyclopedias, to dig out more information.

The great disadvantage of distributing newspapers this way is probably habit. The newspaper, like the book, is a convenient package for information and it has lots of uses after we are done reading it. But will we give up the morning paper at the breakfast table, the ease of passing sections of the large weekend editions among the family? Possibly, I think even probably, but not soon.

We often hear it said that a major advantage of newspapers delivered over the air or by telephone is that we can ask for only the sections or subjects we are interested in. Why get the entire Sunday *New York Times* when we want only the book review and sports sections? This, in turn, offends people who like serendipity — coming upon an article they never would have known might interest them, but does. Actually, I doubt it would work that way. I think it is more likely that we could, at our own option, always see an index or a set of summaries of news items, and then pick from these the full articles we want to read. There is no reason why we have to be denied browsing or control, and we can still avoid having the entire paper delivered in full text. The same technique would presumably be used for any other large publication or assemblage of publications. In other words, the news can be filtered but it does not have to be done in a crude, unimaginative way.

Books

Some of the problems of book distribution were discussed in Chapter 9. The actual transmitting of a book is not too much of a problem. If it has illustrations (more bandwidth required), it would take more than a few seconds, but for a package as large as a book, we should be willing to wait a few minutes. I believe the quality of printing, paper, and binding are the main objections. Either we give up our love of a well-made and bound book, or we encourage someone to develop a home-binding kit, or we develop a really convenient electronic book reading device. All are possible, but not likely to happen too soon.

But there is an aspect to book publishing that is beginning to show major signs of change, and that is customization of textbooks as mentioned earlier. Again, many university instructors are now using collections of published papers or chapters from various books, quickly bound into a sort of workbook instead of a conventional textbook. This approach, called *custom publishing*, gives background readings specifically tailored to a given course. The instructor brings to the custom publisher copies of journal articles, his or her own notes, parts of books, or other materials. The publisher assembles them and handles the copyright permission (or is supposed to). The result is a book that contains what the instructor wants, that does *not* contain material extraneous to the course, and that is up to date. The students do not pay for anything that is not going to be used in the course. The McGraw-Hill Book Company has created a system called PRIMIS[16] that consists of a database of McGraw-Hill publications, mainly books, from which an instructor can easily compile a custom publication. This could become the prototype of the standard way to produce university-level texts, especially if material could be exchanged among publishers, giving instructors a wider range of choices. Instead of text authors writing entire books, they might begin to specialize, writing only a few chapters in their own specialities, then licensing their use in custom-assembled books.

Selectivity of Access

If a journal were published electronically and distributed over the Internet, articles could be posted before they were accepted, and tagged as such. Then, when fully accepted, they could be registered as accepted and formally published. Meantime, readers could find out what's in progress and get copies of important papers immediately, not waiting for printing and mailing. There are some downsides even to this. How would such publication affect or be affected by the sometimes fierce competition in science, and even fraud,[17] which happens in the best of sciences? Then there is copyright. It is an issue still in dispute whether publishing an article on the web prior to publication in print form nullifies copyright claims on the print form.[18]

While many users of the web are delighted with their ability to find useful information, others find that it is often not possible to be precise enough in the questions posed, with the result that too much is retrieved, much of it useless. The Dialog information service claims to have 25 *times* as much data as the web.[19] Such a collection would be impossible to scan meaningfully without a query language (the language for posing questions to the computer) that allows for a high degree of precision. For example, it helps to be able to specify dates or date ranges of publication, to indicate an interest in any word that contains the sequence DIETHYL, or to ask how often a name occurs as an author in any of a set of medical journals during the period 1980-89. With a collection like Dialog's, used chiefly by professional people, it would be impossible to get satisfactory results without a language that allows for great precision. Such languages exist but add to the cost of a retrieval service and demand training on the part of users.

In general, whether in the form of journal articles, news reports, books, TV programs, or what have you, the sheer amount of recorded data today is enormous and growing. (I am using *data* here in my formal sense, defined in Chapter 1 — symbols that were not received and understood or whose reception does not change the recipient.) Combine this with the modern pressure to get more work done faster, and we find ourselves faced with a huge problem of selecting what is worth reading and then, of course, the equally huge problem of reading it all. It is very much the same in other aspects of life. When buying a car or clothing today we are offered a greater variety of possibilities than we have ever had before. To make good decisions we have to invest more time in gathering and evaluating data or learn to rely on recommendations of others or on brand names. These selection mechanisms are not yet as fully developed in the information market as they are in others. The rapid development of communications technology for *sending* data, and of technology for *searching* to provide user access, exacerbate the problem in a sense. We have not moved as fast in providing assistance to those trying to exercise some intelligent selectivity as we have in providing the mechanisms for those who either know exactly what they want or just like to browse. One result: often too many items are retrieved with consequent user frustration.

While many users of the World Wide Web are delighted with their ability to browse and find useful information, others find that it is often not possible to ask a precise enough question to be reasonably sure to retrieve only useful information. When the web retrieves 1,500 records in response to a question, it is a daunting task to browse through them and it may not be possible to rephrase the question in such a way as to eliminate most of them.

Another aspect of selective distribution is the immediate delivery of highly specific information to the consumer. Years ago, going to a library to find a known book, and then discovering that it was not there, was a common occurrence and one we simply had to get used to. There was no way to know before the actual search what was going to be on the shelf. One of the most frustrating situations is when the book has neither been checked out nor is at the bindery, but in fact is in the library on a reading table, left by a previous user. No one knows where it is until the staff makes a sweep through the library looking for books to re-shelve. There are readers who purposely misfile a non-circulating book in a library's stacks so as to insure that only they can find it again. Outright theft is an even more serious version of this situation. Even when you find the book, there is sometimes the problem of finding some specific bit of information within it, a fact or a quotation that may not be indexed.

Those of us used to electronic information retrieval have become quite addicted to stating what we are looking for and expecting to see it before us in just a few seconds. This can be done with library catalogs and with all sorts of databases on-line, but cannot be done with the complete text of all books in a major library. That, plus the fact that with enough computer memory electronic books could be used by more than one person at the same time, makes for one of the more attractive aspects of electronic publishing. An IBM television ad shown in Canada depicted an elderly grape grower in what seemed to be Italy, proudly telling his granddaughter that he had completed his doctoral thesis and that this was possible because Indiana University had digitized its library, with the implication that he could then use their materials remotely. Nice, but this is not what happened. Indiana digitized some music in its music library, not the entirety of its main library. Many libraries have digitized catalogs, but they only help find things. You cannot read the content through the catalogs.

Distribution of printed objects, which are relatively heavy, is expensive and time consuming. Reducing the distribution time almost to zero and the delivery cost considerably is an attraction of electronic publishing. But our present, slow system is versatile. We have a lot of choice over what we receive. (OK, not over junk mail — whose electronic counterparts are just as ubiquitous and irritating.) It is not necessary that we give up choice and control. It may require that buyers let sellers know, emphatically, what they want.

Chapter 11
Comprehension

Is there an issue concerning reader comprehension when using multimedia or other forms of electronic media? Do computers and television change the way we perceive information? Do they affect our creativity and, if so, in what ways? There is an issue perhaps only because many people think there is, and multiplicity of opinion is enough to make an issue or, as Shakespeare said, "thinking makes it so." Unfortunately, there seem not to be any clear-cut answers to these questions. We have touched on some of the points previously.

Some Thoughts about Measurement

My own feeling is that the questions just posed cannot be answered definitively, at least not in our own time. They remind me of the issues of race and intelligence that get raised every now and then, where the neutral observer, much like the observer at a tennis match, watches the claims being batted back and forth, but in this game no one ever "wins." Biologists, geographers, and anthropologists keep telling us that race has no precise definition, and most psychologists and other behavioral scientists tell us the same about intelligence. How, then, does one measure the effect of race upon intelligence, whether to prove or disprove the dependency hypothesis? Some protagonists do the measurements anyway and have even taken to spicing the arguments with comparative measurements of body parts.[1]

So far, the arguments about the effects of media have not been either so heated or, in my opinion, so ridiculous as the race debates. The point is that the latter debates persist in spite of what some of us think as overwhelming arguments on one side. No one, to my knowledge, has ever produced a "fair" test based on a true random sample of the races that has demonstrated the superiority of one racial group over another. Much of the same situation exists with regard to the societal impact of new media. It might take a generation or more to see the net effect. By then the social milieu in which the study was done would have changed. We do not have parallel worlds, one with and one without the new media, that would enable us impartially to study the differences the media make.

How does one measure the effect of style or organization of instructional material or literature on the development of ... of what? *Intelligence*, in the sense of reasoning skill (a most difficult concept to define)? *Knowledge*, meaning how many or which facts a person retains in memory? *Ability* to perform well in institutional settings, such as schools or businesses? Since we can only control to a limited extent what a person reads, hears, or watches, and since we do not have highly reliable ways to measure these factors, I'm afraid we are going to have to settle for discussions and arguments, not proof, for some time to come. A caution to the reader: Do not become beguiled by the literary or computational analog of the body-part measurers. Make sure when you read of a conclusive proof of the superiority of the new over the old or the old over the new that what has been measured really means something to you.

Another way to look at the three factors is that they are concerned with *comprehension, retention, access,* and *decisiveness.* Comprehension means our ability to understand what we have read or become aware of. Tests of reading comprehension are common in schools. A student reads a text and is asked questions about its content that are intended to measure understanding and memory. But there is a hazy line between understanding *what* was said and understanding the *significance* of what was said. Retention is the ability to remember what has been learned, after some elapsed time. A common way of criticizing instructional material on new media is to question students' ability to retain the material over time. Access is the ability to find information. Skill in doing so could be thought of as an application of retentiveness – one remembers not the fact but where the fact is stored, or how to find it. As Samuel Johnson said, "Knowledge is of two kinds. We know a subject ourselves, or we know where we can find information about it."[2] Finally, by decisiveness I mean not just the willingness to make decisions but the ability to make good ones. These definitions overlap to some extent. All the factors are difficult to measure reliably.

When we evaluate or work with certain media, we often try to evaluate them in terms such as these. Unfortunately, it is not always possible. Five years after I left high school and a French teacher I did not much admire at the time, I found myself in France and often the only one in my group who spoke any French. I amazed myself at what I could remember after having been a mediocre student of the subject, at best. Either I learned more than my teacher or I thought, or I retained more than I expected of what meager amount I learned. So, dear Miss Phillips, I may not have done well on the tests, but I performed well on the job. Educational testing rarely can be done over such long time spans.

Another way to look at the issues of evaluation is to start by asking what we want from new media. The designer of a new machine must consider what it is intended to accomplish, not just how it works. What do we want from the new media? It is as unfair to ask of the new that they solve all educational problems as it is to blame existing media for all our existing problems.

A Parallel to Mathematics

When I was in high school, the computer, in the sense we know it today, was just coming into existence. These were giant, room-filling machines, hardly the sort of thing one might have in the classroom. There were desktop calculators, but even these were too large and expensive to consider for private use. In our science classes, calculations were done by hand, or with the aid of logarithmic tables.

When I was an undergraduate, initially as an engineer, there was some controversy over using slide rules in the classroom. In engineering classes, we were taught and required to use them, but in, say, biology, some students did not have them or did not know how to use them. This gave us a great advantage and so it became a moral issue.

Much later, when pocket calculators became available, and at a price that allowed almost anyone to have one, there was again controversy about their use in school. Was their use by some students, but not all, fair? Did their use deprive the students of learning how to do the calculations? Did they have to know how to do them if machines were so readily available? Would it not be better to spend the time learning what to do with calculations rather than performing the tedious mechanics?

Today, I find students in my classes who are not quite sure they know what a logarithm is and I have to remind myself that logarithms, themselves, were mechanical aids to calculations, although software, not hardware. Also today, there is an even higher order of mathematical software that can do more than arithmetic. They can manipulate algebraic symbols, do integrals and derivatives, fit functions to data, etc. These are the things we want to use mathematics for and we can now tell our machines what we want done and they do it. Is it better or worse than in the day when we had to know not only *what* to do but also *how*, in detail? Yes, it's good to know how, but it takes a great deal of time that might be spent learning more theory or application.

To some extent, this is parallel to the question of the use of computers, for writing or reading, or for the use of multimedia to enhance learning. Yes, we relieve the student of the old-fashioned need to study texts. Is that good or bad? It would seem not so bad, so long as we replace study time with learning even more, by use of more efficient learning media.

Linear vs. Non-Linear Text

It seems safe to assume that all readers of this book grew up with conventional, linear texts. Recall that what this means is that the symbols that comprise the message are recorded one after the other, in sequence. This is true of the letters that make up words and the words that make up paragraphs. It is not necessarily true of the ideas that make up a book. Authors can freely jump all over the place in terms of ideas, and take their readers with them. But it must certainly be conceded that printing technology does not make it easy for a reader to jump around in a sequence of his or her choice It can be done with dictionaries, but it's much harder with novels.

It seems to me this issue is very closely tied to the question of the value of or need for discipline in reading. Complex issues require concentration. Authors of linear text have been taught to present material in some logical order, expecting readers to learn what is presented early and use that to understand what is presented later. Accepting the author's order and staying with the prescribed sequence is what I mean by discipline of a reader. It also represents a willingness to accept that the author presumably knows the subject better than the reader and for best results should be followed in the recommended sequence.

I suspect that people who, like me, grew up with linear text and radio, but without television, tend to accept the superiority of linear text and are generally willing to concede to an author leadership on the author's own subject. But those who grew up with television and were early introduced to computers may not feel that way. For one thing, they are more used to short, high-intensity presentations of information, usually calling for less viewer involvement than does text. They are not so willing to accept authorial authority. Are they wrong? A more important question is, Are they any less intelligent, knowledgeable, or accomplished than we are? I doubt it. I think it's a draw. Maybe this is not the fundamental question of the age. Maybe it doesn't matter all that much. Maybe something other than linearity is the major contribution to our learning abilities and accomplishments. This seems to be at least part of the approach to teaching exemplified by Blumberg,[3] described in Chapter 6. In his case, the words would have been in conventional, linear sequence, but the order of reading documents or portions of them could be quite variable.

Uni- vs. Multimedia

Possibly, this is an issue related to that of linearity. The basic question seems to be whether or not we gain from information coming to us simultaneously via more than one sensory channel. Biologically, we humans are easily able to see, hear, taste, smell, and feel at the same time and when we do primitive tasks like hunting, cooking, and eating, the multimedia inputs provide a distinct advantage. We also gesture while talking or singing, and wear costumes when performing before an audience. The simplest and oldest form of *recorded* multimedia is the illustrated text. Here, the two – the picture and the text – are not strictly speaking simultaneous, since the eye can focus only on one or the other at a time, but we can switch back and forth very easily and quickly.

Then there is the motion picture with true simultaneity. In cinema some sounds directly add information to the image. A gun is pointed and a bang is heard. A storm-tossed ocean is pictured and the sounds of wind and crashing waves add to our understanding of what such a storm is like. Other sounds create mood but do not directly present information. The most common form is background music. This helps put us in the frame of mind to receive the message of the pictures and spoken words, but they do not directly portray the image.

Computers also offer the option of true simultaneity. An error in input can be pointed to on the screen and the problem described by spoken words. A computer could display the text of a retrieved magazine article while, at the same time, narrating a reviewer's opinion of the paper. And, of course, anything a motion picture can do the computer can do, too, except to provide a screen that virtually fills the viewer's visual field. But with the virtual reality helmet, even that is coming.

The more puritanical among us will insist on the primacy of text, unadorned with music. The reality may well be that many people will learn more with these media cocktails. Neither my mother, when I was growing up, nor my wife today can understand how I can work with a radio or. I listen to background music and I think I work better that way, at least at writing and figuring, perhaps not at reading. My children do something similar, but they listen to rock music, which to me is an assault on the senses. I can't imagine how they concentrate with that din behind them. So, go figure. I'm casting my vote with multimedia as a potentially good thing. Of course, it has to be done well, and using multiple media does not, in itself make a work good. In fact, that is a phenomenon to watch out for – the hastily compiled work whose main claim to fame is that it is a work in a new medium.

Reading vs. Creating

Picture a small school in the old days. Use your own definition of old days. Possibly its library has one encyclopedia and a few dozen other books. It is in a community where relatively few families have books at home. A teacher gives an assignment to a class to do some kind of research project. The encyclopedia is just about the only book that can serve as a source. The teacher has probably seen every article in it. If any student does the assignment by copying or only slightly paraphrasing the encyclopedia, it is going to be instantly noticeable.

Modern schools tend to have much larger libraries and the cities they are in may have even larger public libraries. About one-third of the homes have computers in them, some of which are connected to the Internet or have CD-ROM encyclopedias. Even without computers, most homes have some books, newspapers, or magazines that might be of use to the student.[4] The range of sources is immense. There is no way a teacher can keep up with everything published. If a student copies or paraphrases the text of a library book, obscure journal, or ancient thesis, it isn't going to be noticed, at least not easily.

Note that I have not used the words *cheat* or *plagiarize*. There is a difference between cheating and honestly believing that the right way to do research is to search, extract, duly credit the source, and integrate materials from several sources. And, particularly in some high schools and even universities, this is recognized as the thing to do. But real research involves interpreting or adding to what others have done. There is supposed to be some originality, whether new information or a new way to interpret old.

If there is so much information "out there," and it is so much work and such fun to find it, is there not the danger that students (and grown-up researchers as well) will tend to be content with searching and finding? Very few readers or evaluators will be able to tell what is retrieved from what is created. And if you get into this habit in your early years, can you ever get out of it? That, then, is the problem.

In an article reviewing a number of multimedia programs for children that encourage them to create original writings or drawings, Gerry Blackwell made these points:

> Much of the value of creative activity for kids [and for adults, too — CTM] is not in the quality of the result, it's in the process of creating....
>
> [T]he essence of any creative activity is generating ideas, the more original the better.[5]

I think the point being made here is again similar to that of Blumberg. Books, TV performances, or what have you that stultify the mind are not good. Such works should lead the reader or perceiver into some original thought or perception. Where I disagree somewhat with these authors is that I do not think it is the nature of books, or even of television, that cause the problem. I think it is the quality of the works, and I particularly agree with Blumberg that many modern textbooks have had the creative lifeblood squeezed out of them by a need to conform with all sorts of dogma. This may be political, bureaucratic, or religious in origin. But there are bad works in all media, and there can be good ones as well.

As I said earlier, I think this is an issue that will separate the true scholars and innovators from what I might call the excavators. Some students will learn that the way to do research is to find what they want (that is, dig it out), copy it, revise it, and submit it for an A. These are the excavators. Others will also search for, retrieve, and copy, but then they will start their original work. They will look for and explain contradictions between established authors, look for what *isn't* explained and wonder why. In truth, the difference between these approaches has always been there, and probably always will be. So, I do not think that the electronic libraries make for more than an extra temptation. They are not likely to be the sole cause of wholesale mental blighting.

If there is a problem it may be the extra load put upon evaluators whose job it is, among other things, to recognize and reward originality and if not exactly punish unimaginative work, at least refrain from rewarding it equally with the really good stuff. This really is not easy. It may be that the danger of overuse of electronic libraries lies here. And our educational systems of late have tended to add more and more to teacher loads, making careful individual evaluation of students' work ever less likely.

The Value of Learning to Search

Several strong proponents of hypertext have pointed out the value of giving students materials and some guidance and letting them work through these and develop for themselves a sense or a model of the subject they are pursuing. Formally, this is known as *constructivist* learning.[6] I share the belief that this approach could be a great benefit to truly creative students because they find the pieces and then they themselves integrate them, creating theories in the process. I also believe that there is great benefit for anyone highly creative or merely average, in learning how to search for and evaluate information.

A conclusion I and others who study information retrieval system users have come to is that many users of such systems simply do not know how to phrase a question, or to systematically search for information, or to evaluate what they find. In this modern world when there is so much recorded information on file and it is so readily accessible, it becomes essential that an educated person know how to deal with this mass. To oversimplify, an average student at the high school or junior high school level is apt to accept any book or article on his or her subject, so long as it is not too difficult to comprehend or in a foreign language (actually another form of too difficult). A more experienced person is going to be concerned with how authoritative this author is, how old the document is, the point of view expressed (or not expressed but implied), and whether the document points out others that might also be useful. Both people might have difficulty deciding what terms or subject headings to search for. Knowing the Dewey decimal classification is not enough, and few of us do know it. We also have to have some sense of how catalogers have interpreted the texts we might be interested in.

The sort of evaluation that helps us find more or better terms from what we were first able to retrieve is not often taught in schools, but the further any of us goes in a career the more important that skill is, and the constantly increasing size of accessible collections of data merely adds to the pressure to learn how to cope with them. A new expression is *data mining,* which is intended to mean digging out, analyzing, and refining useful information from great volumes of data, but unfortunately sounds like excavating.

In a world gone mad with litigation, one of the favorite variations of which is malpractice, imagine a lawyer or physician being sued for not having recovered the precedent or research paper that might have saved a client from jail or a patient from serious illness. Imagine a stockbroker being sued for having missed the bit of information that might have warned about impending problems in a company he or she recommended to clients. Imagine a librarian being sued for failing to recommend a book the patron needed to pass an exam. All these have either happened or been discussed as likely to come soon.[7]

I don't recommend learning to search solely as a way to avoid lawsuits. I recommend it as the mark of an educated and proficient person in this modern world. Peter Drucker pointed out the difference between a person who uses information and one who provides information tools.[8] He was describing the relative roles of the corporation's chief executive officer (CEO) and its chief information officer (CIO). The CEO is a decision maker who must use information to make choices. The CIO's principal job is to provide the CEO and others with the tools – information and information processors – needed to make those decisions well. It's the CEO who is in danger of not being well-enough trained to ask for the right information or to find the useful kernels of it among much chaff.

Another aspect of the reading issue is that there is some evidence that "reading" on a computer is done like scanning of a text, that we *look for* information on the screen but do not necessarily absorb it.[9] This would not apply to looking up a telephone number or something equally specific and identifiable, but to reading a reasonably serious text.

Inspiration vs. Presentation

A subheading over Blumberg's article about the web-based instructional program suggests that the web "can release students and teachers from the tyranny of the textbook – but only if they want to be free."[10] Not all want to be. Not all are able to be. Not all texts are tyrannical. What I think is the most important benefit obtainable from educational material is inspiration. Some people have all the inspiration they need and these are whom I think of as the innovators. They only need resources. They will be self-motivated to make use of them. They will have sufficient levels of comprehension to make use of them. They will know how to find what they are not given.

Not everyone is so fortunately endowed and I refer not to dullards but even to very bright people who have not yet found their bliss. These people need to be inspired. If textbooks are all as dull as the Blumberg article's heading suggests, they are hardly inspiring. (He told me those words were written by an editor, not by him.) But I've seen a few that are, or anyway that inspired me because of the way language was used to express complex ideas that made me see what was not seen before, or at least want to explore more what might otherwise have been a dull subject. These books were written by people who knew their subject, had a command of the language, an interesting style, and a sense of how to organize the material so as to lead a reader through it in a way both logical and interesting.

When we compare the best of multimedia with the dullest of classroom lecturers, the dullards do not fare well. But not all lecturers are dull, even if not all are brilliant and witty. Some are inspiring and I cannot but think I am more likely to be inspired by a real person with whom I can actually talk, than a multimedia course. I think the comparison might be even more direct in the field of children's literature. Children's stories on disc are available. I, by the way, like novels on tape, read by a professional actor. But I wonder about children raised on a diet of recorded storytelling. One of the many skills of a storyteller is the ability to modify the tale just a bit to suit local conditions, perhaps to involve the kids even if just to get them to hiss or cheer at the right times. Maybe to give some special attention to a child who needs it. If you grow up on the Disney versions of everything, have you perhaps missed an important part of your education? Have you missed out on the sense of inspiration you get from a live and audience-responsive storyteller, librarian, teacher, or parent? "Audience-responsive" means interactive, and although I postulated virtual novels in Chapter 7, this sort of thing is not yet available, and not likely at a high-quality level for some time yet. Another perspective on this issue comes from a teacher of music in an elementary school:

Every day I go into my classroom and try to give children an opportunity for first hand experiences: to sing, to move to music, to make their own sounds, to experience the joy of ensemble, to acknowledge and provide an outlet for emotion. From my perspective, most of the vast world of secondhand electronic experience is not an aid but an impediment to my efforts. In addition to restricting learners, polished TV and computer presentations create unfillable expectations. Nothing we can do in the classroom will *or should be* as slick, as instantly satisfying as "with it." When on the occasion we do succeed in making music, I know that I am for the moment fulfilling a basic human need. Even then, I still must convince my students that this is indeed "good," better than watching someone else have this God-given experience. At such times, it seems that all that glitters on electronic media is indeed fool's gold.[11]

Again, this is not a black-and-white issue. Recorded information can be good, entertaining, even inspiring, but it cannot have a personalized human touch. I was inspired to switch my university major to mathematics by an instructor with whom I did not subsequently become close. And yet I can still remember well the lecture that made up my mind to adopt mathematics as my major. It wasn't delivered just to me, nor personalized in any way for me. But it was delivered by a real person, not a movie star or super professor. He was just an instructor, a junior member of the faculty, but now remembered by me as fondly as any other instructor I ever had. What hit me was that *he* knew how to do these things, unlike an actor in some film. Maybe I could be like him. Inspiration.

Chapter 12
Adoption of New Technology

What makes people decide to buy, use, or become involved with new technology? Can it be the clear superiority of the new over the old? Cost? Performance? Ego satisfaction? Persuasive advertising?

I recall two technological developments of which I was what is called an early adopter. The first was the original Xerox machine, the Xerox 914, which came out in 1959. I was not involved in my employer's decision to buy (or more likely to rent) these machines. I was, however, working on a book manuscript when we got ours and having this heaven-sent machine around was a tremendous help to me. Previously, when a typescript page was amended, the options were to retype, paint over the old with that white, chalky stuff, or paste some white paper over the old and type the new text on top of it. Either of the last two options left you with a piece of paper growing in thickness, likely to stick to any one placed on top of it, and unacceptable to the publisher's copy editors. The former option, retyping, was painfully slow for me or required expensive hiring of a typist. Photocopying was liberation and no machine available to me before this particular one was appropriate to the job. So, the switch-over in work habits was instantaneous for me. Xerox won on the basis of both cost and performance.

Twenty years later I converted to the use of word processing when I had to deliver a manuscript to a publisher in camera-ready form, had to type it myself, and discovered on the day I was to mail it out that I had formatted it wrong. It took an hour with the university's word processor to put everything to rights and get the paper in the mail on the same day, for on-time delivery. I decided then and there that I would never again do any serious piece of writing without a word processor of my own.

But I was a very late adopter of color television and compact discs for music. I would never have bought an automobile with cruise control or a cassette player in a car if I had not bought the last year's model one time with these features already in it, and become instantly addicted. Like many people, our family found that playing cassettes during a long trip could keep the then-small children occupied, and I realized how much physical discomfort cruise control could alleviate. These were benefits we had not foreseen when buying the car. Being unforeseen, they did not enter into our decision-making process.

Roughly speaking, we can classify the reasons why people adopt technology in three ways: *social*, whether personal ego satisfaction or the urge to keep up with the neighbors; *cost*; or *performance*. Let us consider these in terms of publishing and the potential switch to electronic publishing.

Social Factors

Much study has been devoted to the question of how people decide to adopt new technology or innovations, which include ideas and methods as well as hardware. Everett Rogers identified five stages in the decision-making process:[1]

> The *knowledge stage*, in which decision makers learn about the innovation and the need for it. *Need*, here, should be interpreted as meaning whether, or the extent to which, there is a need.

> The *persuasion stage*, in which the decision makers consider one possible decision against others, including the one to do nothing. Discussions, studies, or experiences of others may play a large role here.

The *decision stage*, in which decision makers come to a conclusion about whether to adopt or not, or what to adopt.

The *implementation stage,* in which the innovation is put into use.

The *confirmation stage*, in which the decision is reconsidered. Was the right choice made? In my own experience, this one is often skipped in bureaucracies or even families, neither of which like to have to face having made a wrong decision. Privately, many of us wish we had never bought that car, suit, or house. We do not usually like to do a formal and public study of why we made the mistake.

Sooner or later, almost all of us, however conservative, will adopt some innovations. What can be said about those who tend to adopt sooner, or later than others? Rogers classifies adopters as:

Innovators, who are adventuresome, well out ahead of the crowd.

Early adopters, who are "respectable," i.e., not seen as radicals by their peers and superiors, but still ahead of the majority.

Early majority, the center of the crowd, who are deliberate but not immovable.

Late majority, who are skeptical or perhaps extra-deliberate, willing to move but in no hurry to do so.

Laggards, who prefer to stick with the traditional (possibly just because it *is* traditional, not necessarily better in any other way).

These are not rigid categories and any one person may be in a different one for different types of innovations. People who value every possible kitchen appliance in their high-rise condos may own a cottage or cabin in the woods with no electricity or running water. And the reaction from others may vary. Innovators may be viewed by their peers with admiration, envy, skepticism, or disdain. They may be motivated by the reactions they get or by a chance to achieve a genuine accomplishment. Innovators are not likely to give much consideration to the possibility of failure.

I think "laggard" is an unfortunate word. It seems almost to carry some negative moral judgment. Those zero-tech cabin owners live the way they do because they like it. Similarly, people who like to read a lot may be devoted to well-made books but be perfectly willing to work from computer printouts or displays at the office.

On the other hand there are adopters who adopt only after need is clearly proven. That is why bank checks have had account numbers printed in computer readable characters since the 1950s. The banks, not always our most forward thinking of institutions, were becoming swamped by the number of checks in use, and manual processing was becoming impossible. They *had* to adopt, and adopt a brand-new technology. Then, there are some laggards who are not just reluctant but scared witless with the thought of what certain changes might bring. I have seen many examples of this phenomenon, but a combination of libel law and good taste prevents me from telling about them.

Much of the strategy for encouraging people to adopt, otherwise known as advertising or salesmanship, is to show a benefit and downplay the risks and costs. Adopting is different from merely buying. It means buying something different from what has been bought before. Advertisements for beer and cars stress a brand name, a model or type, and social benefits (sex, prestige). They do not particularly aim to convince non-drivers to become drivers or wine drinkers to switch to beer. They want to create and maintain brand loyalty.

Most of what we have seen and heard about electronic publishing to date has concerned innovators. They are the ones who surf the net and who bought the early multimedia computers. They love innovation for its own sake. How, though, are publishers going to convince the broad middle of the book-reading market to adopt? They will need to offer more, whether tangible or intangible, than just the fun of adopting.

Cost

It has been my experience, which I cannot support with statistics, that major technological developments that are supposed to save money usually cost more, at least initially. This was especially true with computing in the 1950s and 60s when we were still largely developing custom software for many applications such as payroll or information retrieval. By the 1970s and 80s this began to change, at least in part because there came into being good generic software that could be customized easily to produce a system for most needs. In terms of electronic publishing, including distribution, I believe most cost factors point *toward* adoption, not away from it. Possibly, this is true because many costs are unknown or unforeseen at the decision-making stage. Here are some specifics.

Paper

In 1994 there began a rather steep rise in the price of the types of paper used for books and newspapers. This induced something of a panic in the publishing world. If the trend, shown between the arrows in Figure A9 had continued, there would have been ample cause for panic.[2] But it did not continue, and people in the paper industry told me that such fluctuations are not unprecedented in that industry. The general trend since 1990 is up slightly. But the very fact that we had this spurt lasting over a year may justify a feeling of uneasiness by publishers and, even if the price has come down, they may worry about another upshoot. Hence, while the situation is uncertain, it may be contributing to publishers' wanting to at least be ready for electronic, paperless publishing.

Computers

Since personal computers became really popular in the early 1980s, their price has hovered around $2,000. But there have been enormous increases in speed and storage capacity, which have given us great increases in performance per dollar. Their size and weight have tended to decrease, although not so dramatically. We have gone from desktop models (still available) to portables which originally meant luggage size, to laptops, which fit easily into a briefcase, to personal digital assistants (PDAs), which fit into pocket or purse. I know of nothing to indicate that the trend toward increased performance will not continue. The size is limited by our current need for a screen, our habit of liking book-page-size pages, and the need for a keyboard for many uses. Possibly a virtual reality-like device would work, one in which images are projected onto a device worn something like eyeglasses. Someday, devices might even project directly into our brains. Such devices would be expensive, in today's terms, expensive enough to discourage large-scale adoption. But someday such equipment will probably come within the economic reach of most middle-class people.

Distribution

The big part of the cost of distribution of print media is the physical toting of rather heavy objects from printer to publisher to retailer and then to the home or office of the consumer. If we can find an acceptable way to distribute electronically there is little question that this *could* be done for much less. Negroponte makes this point – it is cheaper and easier to move bits than atoms.[3] Bits, in this sense, are electromagnetic pulses and atoms are pieces of matter that have weight and take up space. Books or equivalent documents contain a great many of one or the other. The bits are far cheaper to store and move.

Until now ("now" is early 1998) electronic transmission of even large files on the Internet has been cheap – dirt cheap. In fact, to many of us, it has been free. This simply means that our employers or school administrations have bought the service and do not charge it back to individual users or even individual academic departments. It is part of the overhead. Just very recently, there is evidence that some of the sleeping giants have awakened. In Chapter 10 we told the tale of Bell Canada's entry into the Internet market and its attempt to raise its prices to independent connection services, followed by the Rogers cable TV firm's entry into the same market. Of particular interest is that Rogers announced that interactive television was not drawing much consumer interest and that here, on the Internet, is where the big competition between telephone and cable would take place.[4] This seems to be a case of innovation (a form of interactive TV) not being adopted by enough customers, making its potential provider look to a different market for revenue. Another aspect of this competition is that all the telecommunication services might merge in the near future anyway, so there might still be competition among service providers but not between modes of service.

This interest by the big guns could mean several things. The first is that there's gold out there and the big miners are not going to let the little guys get rich even just by picking up gleanings, i.e., reselling existing communication services at a bargain rate. Users may have to start paying more, but will likely get faster and more reliable service in return. What about the future? Will the combat among giants of cable and telephone bring prices down, or does their interest imply that prices are going up? We do not know now and really cannot know for sure which way it will go. In a truly free market, competition would drive down prices. But is this a free market or an ever-shrinking oligopoly?

Assuming that the cost of electronic transmission continues to be low by comparison with the shipping of books and magazines, we still have to consider what happens when the work reaches the consumer. Today, we either buy a book, and then it is ours to keep, or we borrow it and then we must return it to the library or wherever. Those who believe in permanently cheap transmission costs will say that it makes little sense to go to the trouble of possessing a copy of a work when it will always be available cheaply on the net. This could take some getting used to for bibliophiles, but not necessarily for other readers. Authors and publishers will have to do some serious adopting of new methods of selling information.

The, or anyway an, alternative is to download the work onto a disc or even onto paper locally. In the interim, while we wait for a computer reader that is really competitive with the printed book, we might want to do just that. I doubt many readers would be happy with "books" printed on 8½-by-11-inch paper and bound in loose leaf binders, or stapled together. So we have to get back to the question of how to print locally and nicely and to bind acceptably. There are computer printers that print on both sides of the page with excellent typographic quality, but there are no really good cheap binders for home use now on the market. "Good" here means competitive with bookbinding. Possibly, libraries, bookstores, or Kinko's could provide the service at a modest price and that may both satisfy users for a decade or so and provide revenue for bookstores threatened by direct transmission of books to users while the technology sorts itself out. An intermediate solution is to download to discs, but that, while solving the recording problem, brings us back to the current lack of a satisfactory electronic book reader.

Why be concerned with local printing? Why would people who have become accustomed to finding information through the Internet want to print it instead of storing it on discs and reading it by use of the computer? There is no unanimity in answer to these questions. I think the manner of reading is too established a habit to break easily, and that brings us back to the question of whether or when the hardware developers will build us a really good reading machine. Such a machine, if inexpensive enough, could have a big effect on readers' adoption of electronic documents.

Another aspect of distribution is simply that of getting the desired item into the hands of the person desiring it. Trade book publishers, in my experience, have not been too adept at this. Their forte seems to be to deliver books in bulk to wholesalers. Although some solicit orders to be placed directly with themselves, it is rare that the book will be shipped the day of the order, or the next day. But this is an example of the kind of task computers are best at. Take the orders and, once a day or even once an hour, the warehouse can be given a set of orders, with items in sequence according to how books are stored, together with mailing labels. So why does it take weeks? We discussed the tremendous advantage of direct delivery of specific items of information in Chapter 10.

Governments as publishers are enough to make a die-hard capitalist or anarchist out of anyone. The Canadian government clings to a system of crown copyrights, meaning that the Crown (government) owns anything published by the government. In the U.S. the Government Printing Office has first refusal on many if not all publications of the federal government, but clearly does not see itself as being in business to serve users. They are content with small press runs and distribution to a relatively few bookstores and to depository libraries. But in these days of so much information on-line, having a single depository library in each state (a few have more, a few have none) is not satisfying. If the object is to get information to users, at a profit or avoidance of economic loss, electronics could do the job better.

Competition

What has competition to do with cost? The economic tradition is that free-market competition leads to lower costs. But are there free markets in communications and publishing? Publishers have in the last decade or two been on an acquisitions binge, buying one another out, building ever larger agglomerations. While this does not necessarily reduce competition, because the huge corporations are still supposed to be competing with one another, it can eliminate small, innovative independents. Real innovation is not often a big company characteristic. Recently, many companies have followed a merger or acquisition with downsizing of staff. The increased pressure on the remaining employees can negatively affect quality.

In telecommunications, in the U.S. we have seen the breakup of AT&T, and in both the U.S. and Canada the introduction of competition in long-distance services. It leaves a tangled situation because long-distance rates did come down but local calling and maintenance rates did the opposite. As local phone companies raise the price of local calls, this will be a big blow to computer users. And now, the long-awaited battle between cable and telephone seems to be joined in a market different from where it was expected, but right in the middle of the market that involves electronic publishing.

I am not going to try to predict the outcome, but I do want to stress that competition in telecommunications could have a big effect on the cost to the consumer of Internet services, including downloading of texts or reading entirely on-line. We have come to think of this as free. Cost has not been a major factor in most people's decision to adopt use of the Internet so far, because most users do not pay for it directly. But the one safe prediction is that this situation won't continue for long. Indeed, it never really was free, it just seemed to be, for many of us.[5]

Performance

It is an unhappy fact of life that it is usually very difficult to predict the performance of a technological device or piece of software. There are producers with enviable reputations: IBM, Maytag, Western Electric, Borland. But everyone makes the occasional lemon. In hardware, two items may be identical in design, produced at the same plant on the same day, but one may contain a defect the other does not have. In software, this is extremely unlikely. Usually, if there is a fault, all users encounter it. Further, performance means different things to different people. Automobile advertising often emphasizes rapid acceleration from a standing start or ability to drive over roadless mountains. But how often do car owners actually use these features? My own number one requirement for a car is that it keep me away from repair shops. Information about its ability to do so is hard to find. The best we can usually do is to find how well last year's models did. So, at least part of the problem of unpredictability of performance lies with the producer, and part with the consumer's inability to state what is wanted.

In terms of publication-related information systems, I see three major performance factors: *accessibility*, *cost*, and *quality*.

Accessibility

Accessibility is, in effect, distribution seen from the consumer's point of view. How do I, as a consumer, get *to* the work of interest or get it to me? We are most of us used to the combination of libraries, bookstores, and newsstands that are the common distributors of printed products. The daily paper is readily available in a great many locations, including in the midst of our front-yard hedges, but yesterday's paper may be impossible to find anywhere but in a library. Very large libraries give an impression of having everything, but no place has everything. In fact, the great advantage of the Internet is that if we truly want to search *everything,* this may be the closest we are going to get without the expenditure of a great deal of time and money to travel. Remember, only a small fraction of everything in print is to be found on the Internet at this time.

Bookstores used to tend to specialize, either by subject or type of reader, but we are now seeing a new breed, the super bookstore. Such a store will probably have most of the current fairly popular publications as well as some classics, and even some obscure ones. There has been concern that they may want to carry only best-sellers, somewhat like the radio stations that want to play only hit songs. How can a new song become a hit? Not by way of the hit-song specialist station. But even if they do not specialize too much in their stock, they may fear that super bookstores are not as likely as the specialized stores to have staff who are highly knowledgeable in a special area. In fact, it could be the other way around. If the "big boxes" really try to please customers, they may be willing and able to pay the cost of trained staff.

So, today, if we are looking for some printed work that is a bit off the track, it can take quite a bit of searching in new and used bookstores and libraries. Today we also have the Internet and its World Wide Web, now including the catalogs of major libraries, but *not* the text of very many books. I would say that most people who use the net assume that we shall one day be able to find at least a catalog entry for any book or journal that has been published and even the full text of most of them. The entry of the full text of all those publications is something else. Over and above the copyright issue of digitizing them and making them publicly available is the sheer amount of work required to do so, storage space required to keep them, and time required to search them. Lesk reviewed several major library projects in which printed books were converted to either character or video representation.[6] He reports figures of around $30 to 40 per book for video-type representation, and $120 when the book is scanned and the images converted to characters. When a valued book is near extinction, such costs can be well worth it, especially since copies of the new version might be sold to other libraries and the copyright has probably expired.

In the lifetime of people reading this book, we are likely to remain somewhere between the situation of only a few years ago when we had to visit libraries to search their catalogs, and the world of the future when anything in "print" is available from our highly portable computer connected to a global cellular system capable of transmitting voice, data, and video.

Customization

Possibly the greatest benefit of electronic publishing, if we know how to take advantage of it, is the opportunity to customize works for the individual reader or small group, such as a school class. A work written in the hypertext mode can use information about each individual reader to recommend to or even select for that person a path through the material. The occasional text of today does that, with such a statement as "Readers with a knowledge of ___ may skip Chapter 6." But how much better if we could give the reader what he or she, individually, needs in the way of background, maybe even bringing in text from another work to do it. The only alternative today is for teachers to prepare reading assignments for each individual student, a heavy load of work. The computer offers the possibility of using the constructionist approach to education and it seems to be a reason for adoption of computers by many schools. Actually, I think the constructionist approach could be adopted without computers. Give students an assignment and a host of sources. Let them dig up information and integrate it themselves. But the educational world has tended to see the two as highly connected – adopt the computer first, then adopt constructionism.

Quality

There is not much unanimity on the question of the cost of quality and the effect of electronics on quality control. Some print publishers would, or once did, prefer death to consciously publishing a text with an error in it. My own experience has been that children's book publishers feel that way still, general trade book publishers vary, but few would prefer the noose to the no-no, and adult text and professional book publishers know there will be errors and assume their readers are mature enough to cope with them.

Finding errors in multimedia or hypertext works is much harder than in print. The very fact that each reader may come to any segment via a different path from other readers makes it hard, perhaps impossible, for an editor to see the work as all others see it. At best the cost of editing has to go up. By how much, I cannot tell. Will this lead to less editorial review in the future? I cannot tell that either. I hope not, but I harbor a suspicion that it will. On the other hand, I recently found typographical errors in the *New Yorker* magazine and a book published by Alfred A. Knopf. (My hand trembles even to type such a statement.) I have long since stopped counting them in the *New York Times*. Perhaps this trend is the fault of the times (note: lower case *t*), not the electronics.

Will uncertain quality affect adoption? My puritanical instincts would like to say yes, but my experience says no, as long as the quality defects are minor. Some years ago I worked for a database search service – a sort of weblike service but not free. The databases were licensed to us and royalties were paid for their use. We provided the computers, communication, software, documentation, and user training. Now and then, errors were detected in a database by our customers. If so, they might tell us about them and we, in turn, would inform the database producer. We were in a position like a bookstore owner whose customer finds a mistake in a book. The shop owner does not own the rights to the book's contents, hence has no legal right to change them. At most, the owner can notify the publisher. In my time, we got more content error complaints from Japan than all other countries put together. And those who wrote in did not tend to use the tone of a friend who had found an error and was helping us get it corrected. They tended to demand that it be fixed – a different attitude toward quality or its lack, also seen in automobiles. I think North Americans, for better or worse, are more tolerant of error, and far less likely to let it affect an adoption decision or choice of product.

I believe the key points in regard to the process of adopting new technology are that: (1) it is generally an unforced decision while the technology is new; (2) the consequences are rarely completely foreseeable; and (3) once enough people make the adoption decision and the technology becomes standard, others may be left with little choice but to adopt.

So many people now use Windows software that it may no longer be a true option for users of the IBM-type personal computer. So many people use e-mail at work that, in many fields (e.g., university teaching and research), a nonuser is at a distinct disadvantage. If enough primary and secondary schools adopt and teach the use of computers and reduce emphasis on reading and composition to make room in the curriculum, we may be unaware of the consequences. If secondary schools and beyond do not teach computer use, they may be severely disadvantaging their students. Finally, if many publishers rush to convert to electronic forms, the results may be .... Unfortunately, it's something of a lottery. We are not very good at predicting effects of such decisions.

Chapter 13
Markets

Books, telephones, and televisions have all been with us for anywhere from a generation to half a millennium. The markets the people or organizations who buy or use these media are fairly well known, although there are some variations in what we "know." Every North American author of a scholarly book "knows" that 3,000 libraries, at least, will buy the work. But, somehow, sales of scholarly books keep right on averaging at or below 1,000 to 1,500. That is what the publishers know. And authors feel that the reason why at least some of these books sell so few copies is because the publishers never know how to market our books. I have heard this said by nearly every author I know. If each of our books got a full-page display ad in the ten leading daily newspapers, we are all convinced sales would be much better. Of course, publishers would go bankrupt and the ads would do little good because every book would be advertised the same way. Put more reasonably, publishers do not always know exactly who is buying their books or who might benefit most from them. However much may be known, there are highly unpredictable aspects of the publishing market.

What Is the Market?

For all the doomsday prophesies about the end of the book, book titles in the United States are being produced in increasing numbers. Figure A7 shows the number of newspaper, periodical, and book titles published in the U.S. between 1880 and 1990.[1] The number of titles keeps increasing, but if this continues to increase and if total unit sales (actual number of books bought) do not increase commensurately, it would mean that average sales per title must be going down. Hence, the traditional publishers really are feeling the pinch even if industry-wide sales are up. Like many sets of statistics, those about book production and sales have to be taken with some caution, but they seem to indicate that this medium has not disappeared nor even declined significantly. What we do not see from these data is information about who is buying what, how many units an average buyer buys, etc.

Newspapers, while declining slightly in number, are generally holding their own. The industry is not shrinking in total production. This is illustrated in Figure A8, which shows the circulation of daily papers and amount of newsprint paper consumed in the U.S.A. from 1970 to 1991. The newsprint data show that not only are the titles holding fast but the size of the papers (number of pages) is, too. While dailies and newsprint consumption are down, just slightly, the number of small weeklies seems to keep growing – at least in my neighborhood – but national figures on these are hard to find. The combined book—newspaper data show a general growth in reading material produced and presumably read.

This same point, about the failure of books and newspapers as well as live theater and cinema to disappear in spite of the growth of electronic competition, is made in an article by Simon Jenkins of the *Times* of London. U.S. cinema attendance data, not quite so optimistic, are shown in Figure A13.

. . . [T]here are now more hardback books published each year, more novels written and more sold. The last Publishers Association survey of book sales showed a five percent real rise in revenue in 1993 over 1992, with real-term rises of one to two percent every year for the past decade. There are more bookshops in the high streets of Britain than ever before. There is even one in Trafalgar Square. Twice as many people go each week to the cinema as did ten years ago, despite the boom in videos. There are more live music and drama performances on stage in London than at any time

since the 1950s. There are more concerts.... Of course, many of the activities
lose money and their spokesmen have a professional interest in talking
them down to get more subsidy, but cultural Armageddon stubbornly refuses
to arrive.[2]

Jenkins makes the point that cinema attendance has not gone down in
spite of the creation of a new medium the film recorded on videotape for
playing on a VCR.

Telephone usage is growing, helped by a big boost from the computer
industry. Television is thriving, or must be when you see how much investors are willing to pour into the medium both to advertise products
and to change its technology. This in spite of a recent Canadian Broadcasting Corporation program called *The End of Television*.[3] The end referred to the end of the use of the present type of TV receiver, to be
replaced by a computer. And, since any program coming via computer
can be stored and called up on demand, the old idea of broadcasting
(every recipient gets the same program at the same time) may indeed
change. Of course, the VCR has already accomplished some of this but,
at least among my acquaintances, watching prerecorded movies tends to
exceed home recording as a use for a VCR. But I did not hear anyone on
the CBC show suggest that recorded shows or works would no longer be
created or produced. There is some suggestion that using a computer is
beginning to encroach on children's TV watching time, but not necessarily that what is watched is of any higher quality. Some CD-ROM producers are suggesting just that, but there is junk on CDs as well as on broadcast TV. However distributed or shown, the popularity of broadcast medical and police shows seems well established and durable that is, their
popularity in the market is among the "known" facts.

What about the market for electronic publications? We do not yet have
a generation or even a decade of experience. There is no *New York Times*
best-seller list of CD-ROMs or of hypertext novels. (There are lists but
they do not get the readership that the *NYT* list gets.) An entrepreneur has
not got a lifetime of experience upon which to base a decision whether or
not to invest heavily in an electronic product unless the decision is to do
so in order to explore the market. We seem to be now in a phase of primarily tentative probes to see what this market is like or, in a few happy
cases, the first harvest of results from having discovered a truly new
market. The Toronto *Globe and Mail* reported that the market for CD-
ROM books is virtually at an end.[4] *Publishers Weekly* offered a some-

what more optimistic view.[5] Reference works continue to sell, but not much else does. This does not necessarily apply to games or cinematic presentations, the latter just beginning to be delivered on the new DVDs. What the cinema-by-computer market will turn out to be we do not yet know.

The Market Now

As just suggested, the market still seems largely exploratory. The encyclopedias have been available for several years and have generally been well received. The *Encyclopaedia Britannica*, which pretty well is at the top of its line, has recently developed two electronic forms. One is a CD-ROM base priced at around $150. A second version is on the web on a subscription basis at about the same price, per year. The purchase price for a printed copy, by comparison, is $1,599. The *EB* is hardly an average publication in any sense. The *Oxford English Dictionary*, another case of a top-of-the-line publication, has been published in CD-ROM at a price over $600, clearly aimed at libraries or the occasional really devoted individual. The *Canadian Encyclopedia* came out in 1995 in a new CD-ROM-only form, selling in the $70 to 90 (Canadian) range. It is a specialty publication, not attempting to serve as a general-purpose work. It has sold very well, as Canadian publications go, while its paper predecessor did not return a profit to its publisher. Microsoft's edutainment CD-ROMs are critically well received. These are not quite books and not quite movies or TV shows and so it is hard to tell what, if anything, they are displacing. The National Geographic Society, publishers of the *National Geographic* magazine, have made several CD-ROM and web products. They feel they are beyond experimenting with electronic publication; they are *doing it*. Both the *National Geographic* and *Encyclopaedia Britannica* organizations see their new products as developing new markets, not competing with the traditional paper publications, which they have no intention to discontinue. Their attitude is: new media, new products.

Hypertext novels seem not to be a major force in the fiction market, but they have an enthusiastic, if small, following. Bill Gates, as head of Microsoft, surely one of the world's foremost proponents of multimedia, suggests that although new forms of fiction are being developed, using electronics, "linear novels and movies will still be popular."[6]

DISCIS Knowledge Research, Inc. enjoyed a brief corporate life as a major producer of primarily children's multimedia works. Their CEO, John Lowry, sees the multimedia market as illustrated in Figure 9. The two axes represent engagement and learning potential.[7] Engagement is, in effect, successful interaction. It is the ability of the software to hold the user's attention. Learning potential is an appraisal of what the software is capable of teaching. Both would be different for different people, of course. Neither can easily be numerically measured, but we can recognize that some software is more engaging than others, or has more to teach than others.

The three shaded areas in the figure represent video games, movies and television, and text material. Of course, any of the three might entertain and any might instruct, but the third category is intended to depict works that are primarily for teaching while both the others will have some primarily entertainment component. Games are seen as high on the engagement scale, low on the learning potential scale. Text material is usually the opposite, and movies are in between. Some movies or videos can instruct and they usually can hold an audience reasonably well. The area in the lower left-hand corner represents a market area in which no one would buy anything. Any work falling here would be seen as too lacking in both engagement and education to be worth the buyer's while.

The three lines going roughly from left to right depict typical buying patterns of three groups: educators, parents, and end users (those buying for themselves). Members of a group would typically buy to the right of or above their line. Educators will buy mostly text material, some movies because these may have instructional value, and even a few games for such purposes as familiarizing students with the use of computer equipment. Parents will typically buy more video games and movies and less text material than educators. End users tend to like the games and movies overwhelmingly.

Lowry believes the producers must strive to move their products toward the upper right-hand corner of the graph: works that have some educational or social value yet engage the user.

The Buyers

Who buys electronic works? I do, but I'm in the business and I buy them mainly to see what they are like. Lots of other people with computer experience or who really want their children to acquire that experience buy them. You can read a book about computers without owning one, but

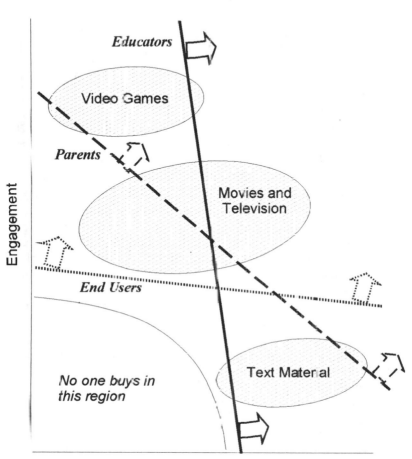

Learning Potential

Figure 9. A representation of the multimedia CD-ROM market. *This is based on a model developed by DISCIS Knowledge Research Inc. The two axes are* engagement *and* learning potential. *Engagement is the ability of the software to hold the user's attention. Learning potential is an appraisal of what the software is capable of teaching. Between the lines we see represented video games, movies and television, and text material, and the area in which no one would buy anything. The three lines or curves depict typical buying patterns of educators, parents, and end users. Members of a group would typically buy to the right of or above the line shown.*

you cannot "read" an edutainment disc without a computer. Plainly, the market for electronic publications is a subset of the market for computers. That is important, because while the computer market has continued to expand for nearly 50 years now, these machines are not yet universal. They are not in as many homes as TVs or telephones.

Figure A13 shows the number of various electronic devices in use compared with population over the period 1940 to 1995. The lines are virtually parallel, suggesting that these measures are all growing together. (The cinema data are from two different sources, which may account for the sharp drop between 1965 and 1970.) They further show that there is near saturation of the market for radio, TV, and telephone, and computers are headed toward that saturation point. Saturation does not mean no more products will be sold, but that almost everyone in these countries has one of the devices available. Future sales will depend on replacements and on multiple units in the home. Note also that this graph is plotted on a logarithmic scale, which makes differences near the top of the graph look very small. There are not actually as many TV sets as people in the U.S.A., but there are radios, TVs, and telephones in over 90 percent of North American households.

There is much room for growth before computers come near the degree of home market penetration these other devices have achieved. The family first has to make a decision to invest about $2,000. Then they can consider electronic books. Yes, they will get a few free discs if they buy a multimedia computer and this is an excellent way for publishers to begin growing a market. If a publisher can break even now, with computers in about 35 percent of North American homes, the growth of the electronic publication market is almost assured as that 35 percent approaches 100 percent. New types of computers and of telecommunications systems could change these economics. The next generation of home computers, some feel, could sell for under $1,000 and be largely limited to communication and display, with the heavy computing and storage done at a remote, large computer.[8] *Déjà vu* all over again. That's how we did it in the 1960s and 70s.

Figure A2 shows some figures about the United States population and the relative number in the age group of today's high-intensity users, who are about eighteen to twenty-five years old, and those older than that. It shows the percent of the population that was in each age group, on average, during 1970-90 and the projected percents for the period 2000-2020. Note the hump in the projection for ages fifteen to twenty-four. They represent the children of the baby boomers now in the thirty-five to fifty-four group. For investors, there is the happy prospect of a large number of young people coming up, who can be expected to be computer literate when (if ever) they become employed and have income to spare. Note here that all intervals cover ten years except the first (less) and last (more).

The Strategies

We are seeing enough different approaches to electronic publishing to suggest that publishers are not yet agreed on a market definition or strategy. Some are now well established, like what they are doing, and see their market responding favorably to their works. Some nervously admit that things look like they might change, but they do not know what to do. For those old enough, think of the advertising blitzes that came with the end of World War II and the spurt in television receiver sales, the introduction of long-playing phonograph records and players, and now of satellite television and wireless telephone. These were all fairly well-defined products when they came out but they were sold as machines or services, not content. Even the then-new kind of phonograph record was like the old in terms of what it did, but its performance was demonstrably better. Now, we're involved in the sale of unfamiliar contents. One attitude is that there is not much point investing large amounts of money in a product being test-marketed, when you know the form may change before the next edition comes out. One of the great attractions of the World Wide Web is that it does not take much to get a publication "up" on it. And its users are so avid to explore it that little or no advertising is needed. But will the web form of publication last? We do not know.

My experience in talking with publishers is that the closer they are to reference or instructional material or to children's programming, the more inclined they are to publish electronically, at least on a test basis, and their products often do well. Companies that produce primarily adult fiction, essay, or narrative-type books are far less venturesome. Rightly, I believe, they do not see these particular books becoming CD-ROMs, nor do they visualize anyone wanting to read 500 pages of essays on a computer.

When the new $33\frac{1}{3}$ rpm phonograph records were introduced, and later CDs, the producers *wanted* buyers to replace their old forms. Book publishers may want that but neither they nor their readers are ready to make that change in large numbers. For some time, yet, the paper and electronic markets are largely different, complementary, not competing except in the reference area. One way, though, to encourage buyers to begin seriously collecting and using electronic books would be to follow the phonograph record model and produce a number of good books in electronic form, then either give them away or sell them very cheaply. Create the market first, then see if you can function successfully in it.

Basically, we are standing on the edge of something that almost everyone involved senses is big. Telecommunications companies are delighted with the increased usage by computers, but they are hard-pressed by the frequency and cost of technical changes. Computer companies, too, are both delighted and hard-pressed to keep up with changes. But both also face a technology changing very rapidly, eating up investment and constantly bringing new competition into the market. Indeed, Stephen Levy, writing in the *New York Times,* suggested that the whole notion of the 500-channel television system being touted by the cable industry "was a myth created by the masters of the media," that TV would go beyond its role in entertainment and become the center for commerce and information as well.[9] He then quotes John Barlow as saying that the current waves of media realliances (the frenetic mergers in the field) are like "the rearrangement of deck chairs on the *Titanic.*" The iceberg, says Levy, is the Internet. This may be a bit overstated, especially considering the financial power and influence of the media giants, but it is a point of view well worth serious consideration.

Some publishers, seeing this uncertainty, are wary of entering the electronic field. But almost all the large ones seem to be testing something and a few are already enthusiastically operating in the multimedia market. On the other hand, a few have abandoned the effort. Yet, there remains a sizable faction among potential customers who feel that printed books will remain a highly marketable commodity for the foreseeable future; the old technology is just too good and too well-established to be easily displaced. This is not exactly a conflict. There may be ample room for both, as we have learned in the cases of radio, TV, and cinema.

To summarize, what we know is that the current buyers are characterized as adventurerous or as Rogers's innovators, people who like computers and like to explore new products associated with them. A second group, somewhat more cautious, is nonetheless as convinced that the computer is the way of the future, and they will enter the market on behalf of their children if not themselves. These are probably early adopters. We do not know the size of this second market segment. All users of electronic publications have to be computer-oriented if not actually computer literate. Publishers are showing an understandable reluctance to abandon traditional lines that are profitable today in favor of untried products. The key seems to be not to *abandon,* but to *expand* into new areas.

A significant segment of any market related to media consists of those who are buying for children. Although the children may influence these buyers, who may be parents, grandparents, teachers, or librarians, there is much advertising aimed at the buyers themselves. They seem, today, highly susceptible to the pitch that anything computer-related is of inestimable benefit to the kids. I don't agree with that claim, but I do agree that the claim sells merchandise.

Obstacles to Market Growth

We are poised on the edge of the most fantastic media revolution in history, but there seems always to be a "but." I see six obstacles to the development of this market. That obstacles exist does not mean they cannot or will not be overcome. It does mean that they have to be addressed. The first five are: (1) no good computer for reading the equivalent of books and no pressure from publishers to create such a computer, (2) lack of authors with skill and experience in writing for the new media, (3) lack of an established model of what multimedia productions should be like, (4) threat of piracy of network-based media, and (5) inertia. The last obstacle seems to be selectively dwindling: (6) the lack of real substance in the market.

To me, the greatest obstacle to the growth and prosperity of an electronic publishing industry is the fact that the computer industry has yet to produce a good, highly portable machine that is well adapted to reading text and looking at pictures in all sorts of environmental conditions. A related obstacle is that the publishing industry has not done anything about it.

But even if the technology were available and affordable, the next major obstacle would be the lack of experienced authors or creators of multimedia or hypertext works. Here we have a new medium or group of media and we cannot expect that there are experienced and skillful authors just waiting for the markets to mature. Ultimately, it is the market that determines the success of an author. Yes, success is influenced by reviewers, advertising, and publishers' reputations, but no matter what the reviewers say, either people buy the work or they do not. It typically takes several tries before a book author is fully accepted. There are always a few who strike it rich with the first novel, but there are many, many more who never get published or whose books serve only to fill up remainder shelves in the bookstores.

I am convinced that the two main reasons why computer-assisted instruction never lived up to its hype are that the hardware was never quite good enough and very few authors ever became really accomplished at the art. There were just never enough courses written. Hardly anyone who was an author had ever taken such a course as a student. No student ever went through his or her entire schooling taking a series of such courses and then decided to become a professional writer of them.

When people, both authors and their readers or consumers, move over from one medium to another they tend to take their traditions with them. So, radio and television developed originally along theater lines, with scheduled performances well advertised in advance. But some of the best TV has been spontaneous, live coverage of news. Those of us who are not particularly enchanted with hypertext fiction object precisely because it is not like the fiction we are used to. So, the third major obstacle is the lack of a model for multimedia publications. What are they supposed to be like? Who decides? The early success of electronic reference works suggest that a multimedia "book" has to use all the media: text, still pictures, motion pictures, and sound, and must *involve the user*. Users do not seem to want to just passively watch multimedia works. They do that in the cinema, of course, but the show must really grab and hold the viewer's attention, and nonfiction cannot always maintain that level of dramatic tension. In television even the remote control unit offers some user involvement. Probably the main reasons for success of reference works in computer form are that the software can search quickly and that we do not have to spend much time reading once we find the article or entry we want.

The fourth obstacle is copyright protection, or its lack. This seems to lurk like a specter over the nascent Internet industry. It is only slightly less of a problem with discs. It has always been possible to pirate books, pamphlets, or anything in print. And it has been done often. The pirates let the original publisher find the author, nurture the work, edit it, print it, and then if it is successful, typeset, print, and distribute their own copies. With modern copying equipment it is even easier, because the work does not have to be typeset again. The author and original publisher gain nothing from the pirated edition, but they have put in time, effort, and money tc create the original. Still, it takes some effort and investment to make a pirated print copy. With electronic publishing, the effort is quite minimal. If the consumer can download from the Internet, why can't the pirate? Copying and reproducing a CD-ROM is also no great technical feat.

The photocopier has had a big impact on copying of copyrighted material. In a famous law suit beginning in 1971, Williams and Wilkins, a publisher of a number of scientific journals, sued the U.S. National Institutes of Health because, they alleged, the National Library of Medicine (NLM) was infringing their copyright by offering free photocopies of journal articles to persons in the biomedical field.[10] After a number of hearings and appeals, the plaintiff lost, by a squeak. American copyright law had long been interpreted to allow for copying in limited quantities for research and educational purposes, called the right of *fair use*. NLM was not republishing for profit, they were merely redistributing articles from journals they had paid for, for research and educational purposes. But the scale of copying was beyond what the world had been used to. A new medium, the photocopier, led to new information habits. With the number of journals and magazines going on the web today, what is there to prevent wholesale copyright infringement? Perhaps what was once considered illegal will come not to seem like infringement in the future. Can the copying be controlled and held within legal or economically acceptable limits? Yes and no.

Yes: it is possible to require prior presentation of a credit card number or password before being allowed to copy from the Internet. The password is an indication of existence of a contract between publisher and buyer and is, in fact, little different from a credit card number, which indicates a prior agreement between buyer and bank. This is done in all sorts of sales transactions by telephone and computer. It is possible, also, to encrypt the text and to make reasonably sure that only a qualified (i.e., creditworthy) recipient has the decoding key.

No: unfortunately, hackers have shown a remarkable ability to get through computer security systems, so none of these protective schemes is perfect. But then, there never has been a foolproof way to protect a publication from piracy. No wonder publishers are wary.

Inertia as an obstacle does not solely mean resistance to change. Remember from your physics that inertia refers not just to the difficulty of setting a nonmoving object in motion but as well to changing the direction or speed of a body already in motion. Just as the entrepreneurs must

worry about people who simply do not want to change, the people have
to worry about entrepreneurs who create a stampede toward change
whether or not there is a need for or benefit from it.

The question of substance is a touchy one with many people. There
are a number of related questions: Does the Internet or the web carry
enough good information to really satisfy the serious, knowledgeable
user? How does the quality of what you can get on the web compare with
that of the content of a first-rate library, *reference services included?* Is
there yet incentive for serious authors or publishers to use the web as a
distribution medium? Real web *fanaticos* will brook no criticism of content. Yet, I have been more often unsatisfied than satisfied in searching
for economic data for this book. It is hard to find because it is not well
indexed. Often, friends have referred me to a site for specific information, which I found simply was not there. When information *is* there, I
sometimes do not know how to evaluate it. If it were economic data and
came from a U.S. government agency or McGraw-Hill or Dow Jones I
would tend to accept it, not being a professional economist. But I may
find papers written by persons connected with organizations I never heard
of. So today I am among the dubious. On the other hand, I believe there
is more good "stuff" arriving every day, more reputable publishers getting onto the web and bringing their reputations with them. The future is
bright, if the present is still somewhat murky.

The Future Market

In spite of all the obstacles, including inertia, the world seems to be moving
toward electronic publishing, a move that may take a generation or even
more to complete. By complete I mean the near total replacement of
print publications by electronic ones. While much of the reason may
simply be the attraction of glittering machines, the rate and direction of
change seem unmistakable.

Two points that seem clear also serve to cloud the future. One is that
the current market is not large enough to endanger the print market. The
question is when might this happen. The second point is that a new medium has to displace something in terms of its users' time.

When motion pictures came on the scene, some other activity had to
be given up. One such activity was live theater, perhaps not top of the
line, serious drama or Broadway musicals, but certainly vaudeville. I
have never seen a study of whether radio displaced reading. But since

early radio attracted many listeners, it had to have displaced something. The coming of broadcast radio roughly coincided with the coming of the 40-hour workweek. Perhaps this was a case of the creation of more leisure for adults and the introduction of something to consume it. Perhaps children were lured away from active play. With television came something of a dip in movie popularity and the shift of most of the most popular radio programs to the new medium. But both came back, all three now sharing a market. Is there more leisure than there was around 1950? Probably, but not a very dramatic difference, if any. So, what did we give up to gain more time for all three media? Radio today is probably always time-shared with some other activity — homework, cooking, driving — so the modern version of radio has not necessarily displaced anything. TV can be time-shared with a limited number of activities, surely not driving, but yes to homework and knitting. Cinema cannot be shared with anything, especially since modern theaters tend not to have a remote balcony where other forms of entertainment used to be pursued without interference.

I talked with a number of bookstore owners in getting ready to write this book, and this was one of the questions I asked: What do you think the new multimedia computer systems will displace in our lives? None answered the question. All said they did not foresee any. But that literally cannot be. If we were not spending time with our computers before, and now we are or are beginning to, what is it we are not or will not be doing? Perhaps that question alone tells us why publishing, cinema, and TV giants are merging. They may not know, either, and don't want to take any chances.

Another way to look at the future market is by demographics. The current major users of the Internet are young people between about eighteen and twenty-five. Originally, they were mostly male, but now women have caught up. This is not a group distinguished by a large amount of disposable income. Many of them get onto the Internet free, courtesy of their schools or employers. They are probably not the predominant buyers of books, magazines, or newspapers. If a publisher currently supplying a service to this group wants to expand its market, what will they know of the next segment into which they move?

The eighteen-to-twenty-fives will naturally age, get better paying jobs, and begin to spend more money. Being used to electronic information, they will be willing to spend some on that medium. But waiting for a population to age is a slow way to expand a market. I think we know something about other population groups, but the important point is that anyone trying to crack the market must realize that the design of future products and services has to be oriented toward the other people being addressed, not this original group, or not just this group. What does this mean? Let us consider a few groups.

The group currently under the age of eighteen is even more electronically oriented than their immediate elders. But they have even less money to spend. They generally expect things to be provided by adults, but they can be very persuasive, as can advertising aimed at the parents of young children. They will be far more open to nonprint media than will older people.

If we look at older groups, those between about twenty-five and forty-five have much more disposable income than the eighteen-to-twenty-fives, but also have other pressures for spending it. They have children. They have mortgages. They will tend to have reasonable computer literacy, but not as much hands-on experience as the younger groups. They may just want a bit more assistance in using computers than the younger ones, who may feel that being offered help is something of an insult. My guess is that the human-computer interface will have to be different for these people. For example, producers should not assume that everyone is adept at using Windows or even a mouse. For someone who last worked on a keyboard in a high school typing class 20 years ago, that mouse can be a daunting instrument.

Move up an age bracket and we come to those about forty-five to sixty-five. Ah, big money. Children *and* grandchildren. Far less average prior knowledge of or experience with computers. Possibly ready to buy anything for the youngsters in their lives, because they are probably convinced that computer literacy is the ticket to future happiness and success. These people buy a lot of publications. Will they cheerfully switch over? I doubt it will be cheerful. I think they will need an even more sensitive and helpful user interface than we usually get today. In fact, it is a valid question whether they will ever switch over, hence whether it is worth planning products for them. To my surprise, several elderly people of my and my wife's acquaintance, who liked to read, would not consider using a cassette tape player and books on tape, when some infirmity prevented them from much reading of printed books on their own. I have heard several speculations on the reason: resistance to a new medium, the fact that the listener cannot control the speed of reading, and an unwillingness to admit to having had to give up reading because of an infirmity. All three center on a reluctance to adjust to a new medium.

I wish I could lay out the precise specifications for what is needed to appeal to each segment of the consumer market. All I know for sure is that each market segment is different and may want different solutions. Mickey Mouse is old enough to be known by just about everyone in North America, and he may sell movies, T-shirts and wrist watches, but old Mick's appeal is not universal. For whatever reason, different people are attracted by different icons. If you want to sell to everyone, you may have to change the icon. I do not believe the electronic publishing industry has yet found that magic attraction.

Chapter 14
Protecting the Consumer

We live in a competitive world. Companies eat one another up, sometimes fleece customers, and sometimes are fleeced by customers or their own employees. A key role of government is to protect the public and the companies from such predatory activities. Self-control or self-regulation works sometimes, usually when the exercise of such control is seen as the lesser of two evils – do it yourself or have the government do it for or to you.

Our new methods of communication, and of publishing, which after all is but another form of communication, challenge some of the traditional means of protection of consumers and companies, by which the competitive playing field is kept reasonably level even for sellers as well as buyers. But regulation tends to be slow to catch up to changes in the world. We must expect a certain amount of chaos in our new information highway, both in terms of new companies being born and old ones disappearing. Consumers can be hurt by too-rapid change, negating the value of their investments in new equipment and technologically dependent works. For example, prior to 1980 IBM was doing splendidly in the world of hardware with a huge investment in designs, plant, and inventory for

large computers. Along came the PC and although IBM got into that
market and dominated design, they did not for long dominate sales.
Compaq makes more computers in his line and Microsoft sells more
software than IBM. Microsoft, almost purely a software company, is now
more valuable than IBM in terms of value of stock issued. In related
markets, Western Union, the telegraph company, has shrunk to insignifi-
cance, as have several watch manufacturers who preferred mechanical to
electronic movements. We may soon see the electronic-game hardware
people losing out to general-purpose computers configured for multimedia use.

Regulation

The regulation of services by a government is the sort of thing that
generates a lot of political steam. In one view, regulation is intrusion by
the government into private industry, the creation of red tape where none
is needed. Or it is a protection for industry from having to give fair
treatment to consumers. Or regulation is government action to protect
existing industry against unwanted competition. (And what is *wanted*
competition?) Or it is an exercise of the duty of government to assure
that commonly owned resources are used in the public interest, and that
dangerous substances or practices and high-risk trading practices are
properly governed in the public interest. Take your choice of
characterization. The subject is brought up here because most publishing
is not regulated nor is very much of the Internet. And the question is,
should either be regulated?

Purpose of Regulation

Why we have government regulation is covered by the second of the two
descriptions above. In North America and much of the rest of the world,
governments control transportation: the roads, airports, air lanes (through
air traffic control systems), harbors, canals, and rivers. In most of the
world a vehicle must be licensed to be on any of these publicly owned
spaces, such as rivers that are assumed to be common property, protected
by the government on behalf of the people. Not only must the vehicle be
licensed but so also must the driver or pilot. Most of us accept this. It has
been with us all our lives. Regulation is especially tight when the vehicle
is for hire, because it is assumed that the average member of the public is

unable to make safety judgments, such as assessing the air-worthiness of a commercial airliner. Freight carriers are regulated in terms of the rates they may charge, to assure that all customers and competitors are treated fairly. Regulation can be used to assure that markets are equitably shared among competitors, or that existing monopolies or oligopolies are protected.

Governments also control telecommunication. As I have pointed out, most national telephone systems are owned and operated by a government agency. This is beginning to change, but even where ownership changes, there is still a government agency regulating the industry – the United States Federal Communications Commission and the Canadian Radio-television and Telecommunications Commission. For most of us in North America, the idea of the government regulating publishing is abhorrent because the forms of regulation we are used to primarily govern carriage or safety, not content.

But, as Internet use evolves into a major industry, should the prices charged users be regulated? Should the quality of service be regulated? Should the content of messages sent be regulated? In 1996 the United States enacted a bill that, among other provisions, outlawed the knowing transmission of indecent materials on the Internet where it might be received by children.[1] It also opened local telephone service to competition, and required the installation of a so-called V-chip, or device to allow parents to control what may be seen by children in their homes. Since then, the indecency controls aspects have been declared invalid by a U.S. court, which ruling was subsequently upheld by the Supreme Court.[2]

The attempt to control pornography raises the old (but valid) questions of what pornography is and where the government's obligation to govern content begins and its right to do so ends. If they can regulate the content from the point of view of sexuality, can they regulate it from the point of view of politics? What about racist remarks, racial libel, attempts to inculcate children into racist beliefs? What about the same applied to adults? What about sending messages across the Internet about how to build bombs or where or how to plant them? Is it right that our tax supported systems, or systems that use the common electromagnetic spectrum, be used to undermine our own way of life? Or is it the regulating that might undermine our way of life?

These are some of the political questions about regulation. They will not be soon resolved. There is the struggle to preserve the right of free speech, but also the question of whether that right is absolutely unlimited and whether public funds that provide free Internet service in schools and libraries thereby should support racism, terrorism, or unspeakable filth. Of course, and unhappily for the regulators, what is my unspeakable filth may be your art. *Chaque à chacun.*

Some Past Experience

The early telephone systems in the U.S. were regulated because, we are told, the assumption was made that only chaos could result without it. There had to be common standards for connecting one phone to another, originally done directly, later done through exchanges. There had to be interconnect standards for exchanges. These came later than the telephone itself. Eventually the manner of regulation shifted to specifying how different systems should be interconnected, rather than that each city or area should be served by only one company. All this was claimed to be done to protect the consumer. It also protected the companies that were given the monopolies.

As it developed, telephone regulation covered not only technical matters, but also rates and corporate profits. It seemed to work, although telephone companies made bundles of money, which some people resented. I remember growing up with excellent service in the sense that I never remember being concerned with a nonfunctioning telephone until after deregulation. In the early 1980s I bought my first deregulated phone for $7.50. Its bell rang once and was silent ever after. The tradition grew that long-distance revenues, mostly generated from businesses, would be used to defray the costs of local service which were more evenly divided among business and residential use. Local service included maintaining the instruments as well as providing transmission and switching services.

The first couple of major challenges to the old system came from Carterfone (see Chapter 9) and MCI, the latter an upstart company that wanted to operate a high-speed data communications service between St. Louis and Chicago. Both these young companies had the audacity to challenge AT&T's near-absolute control of American telephony and consequently of the burgeoning market for computer-to-computer communications. In effect, they were also challenging the whole system of regulation in the courts. They both won, and regulation has never been the same since. Is this good? I can hardly imagine anyone wanting to go back to the days when you had to have your telephone company's permission to attach a piece of equipment they did not make to your phone or phone line. On the long-distance front, yes the rates have come tumbling down, but the local rates have gone up, as predicted. In my area, not only have local rates gone up, but repair service, formerly free, has now been unbundled and we subscribers are now free to deal with our choice of service company and, of course, pay for it separately. When all telephones were rented from the telephone company, that company had a vested interest in making its equipment reliable since they paid for the service calls. Is it possible that only, or mainly, the biggest commercial telephone users have made any real gains from deregulation?

The electromagnetic spectrum has been regulated because it is limited. There is just so much bandwidth. (Bandwidth, recall, is a measure of transmission capacity. See Chapter 9.) And if more than one transmitter uses the same frequency in the same geographic area, neither would be clearly heard. That's what got the government and international agencies into regulating frequency usage. They treat the spectrum as a common resource, to be used in the common interest.

Now we have to deal with issues such as regulation of content, the protection of local culture and usages from foreign intrusion. None are new. All are exacerbated by the physical ease of transmission or transportation, the sheer volume of material, and the consequent virtual impossibility of examining or monitoring "shipments" at border customs stations.

What, If Any, Protection Is Needed?

Copyright regulation for printed materials had and still has a hard time keeping up with the changes wrought by Xerox. The music industry worries about digital recording because cheap, homemade copies of commercial recordings can be of as high quality as the original. So why buy when you can copy from a friend with no loss of acoustic quality? The same would apply to video recordings. Already, we hear allegations that some video rental stores are renting out illegal, inferior copies of tapes, and that lax or nonexistent laws in some countries are depriving music, book, software, and even hardware producers of revenue from the sale of their products.

Telecommunications technology is changing rapidly and each new wave offering new capabilities seems irresistible to consumers. Some of these threaten the existing industry structure. Not that I want to inspire sympatry for companies that have grown rich with our business in the past, but it is or should be a concern for the consumer that companies investing in this industry be able to afford to keep on improving the product. They will not be able to do this if a new technology comes along *with new corporate players* every ten years or so and knocks off the old players. How should this change be managed? By the government? By the market? By associations of users? Telecommunications carriage is moving from the province of tight regulation, with carriers prohibited from doing much in the way of other kinds of business, to a situation where they are combined into conglomerates with content providers or with services that process data. As such it becomes harder to separate the regulation of the carriage of messages from that of the content of messages. There may be risk for the public in this trend toward amalgamation.

Regulation is not a quick-reaction process. It tends not to keep up with major technological or societal changes. What we probably have to face up to is that regulatory apparatus will *not* keep up with the changes we can see coming. Think what might happen if almost all computing were to become remote computing. What is already the almost frightening societal importance of telecommunication will become even more so for our entire culture. Will it be managed and regulated as we would like it to be? Probably not; it is too inert.

One telecommunication change likely to have a major impact is the introduction of personal communication services, or PCS. PCS uses very high frequencies and provides for a greater variety of types of signals along the lines of existing cellular telephone. It handles voice messages as well as computer-to-computer transmissions, and can use satellites to span very long distances. Combine this capability with computers that could be much less expensive than those of today and that rely upon communication from user to a remote computer to do the heavy computing or data storage, and we have a very powerful, highly portable means of telecommunication in voice, data, or images. This could come to replace the present telephone system, heavily dependent as it is on local installation of wires from home or office to the local telephone exchange. It could also replace the present television transmission systems, allowing us to receive the same programs, but with complete user choice of when to receive or view them. At the same time, what we may be doing is shifting the cost of computing from the price of a PC and our own software to that of a $500 to 1,000 computer, *plus* the cost of communications facilities, *plus* the cost of using the remote computer and its software. It is not possible to predict at this time how much money would shift from any seller in the market to any other, whether the net effect would benefit the consumer, or whether this would result in more or less centralization of industry.

It is also possible that some of the major competitors in the telephone, cable television, and PC markets will fall by the wayside, unable to cope with the financial burden of so much rapid change. For example, if the PC were to metamorphose into primarily a communications device and its price were to fall to half its current level, then consumers would be relying on fewer, larger, remote computers to pick up the computational slack. That means less revenue to the PC producers. Yes, the machines would cost less to produce, but how many companies could watch their prices drop in half, regardless of changes in cost, and not suffer?

Again, I am not trying to stimulate reader sympathy for the "poor" telephone, television, and computer companies threatened by this new technology, but to point out that when heavy investment in new technology is so quickly threatened by even newer technology, it is not necessarily good for any of us. It could result in unwillingness to invest further.

It is unclear what these changes mean to anyone or any industry segment, hence also to the publishing industry. If the cost of a computer really does come down to the price of a good TV receiver and if the industry ever develops a computer good for reading text, this price reduction could accelerate the trend toward electronic publishing.

Since we are now on the subject of regulation, we have to consider the question of whether regulatory apparatus and laws can keep up with the rate of change facing us. They (laws plus adjudication plus enforcement) did all right for us with radio frequency allocation, although there are always complaints from those who cannot get a broadcast license. In telephony, regulation was OK, but in the long run needed much change, when the early strict local monopolies, long-distance monopolies, and bans on "foreign" attachments were eventually found unnecessary and anticompetitive. In the U.S. some of the strict conditions approved by the courts when AT&T was broken up are being relaxed as the "Baby Bells" scramble for new markets and partners.

The switch from black-and-white television to color was done well in the U.S., from the point of view of the consumer, whose investment in black-and-white receivers was protected by regulatory insistence on a compatible form of color transmiss on. Not every country handled the transition the same way. When cable television arrived, the consumer again had a choice of using it or not. The usage of cable does not exclude the preexisting program producers or local producing stations from gain, because they get some income from the cable companies, though there can be hard bargaining over what channels are picked up by the cable operators and for how much. The consumer usually can at least count on better-quality reception and more choice of programming through cable. Since there is more choice, that means that some local transmitting companies must lose some of their audience.

As the "death stars" arrive, they *do* threaten the cable companies because they provide essentially the same kind of service, typically with even more user choice. In Canada, advertising for, but not yet delivery of, direct-to-home TV (DHTV) began about the same time the cable companies decided to expand their offerings and did so in a way offensive to many customers. If you didn't want the new channels, you had to ask for them *not* to be provided. This caused such resentment that the DHTV people made some initial inroads and the cable people had to do a quick reversal of marketing techniques. Then, pirated signal decoders went on the market. These devices enable a user to decode the encrypted signal from a satellite without having to pay the satellite service for use of the service. This threatens to steal revenue away from DHTV, and it frightened operators in Canada into at least temporary retreat from the field. There was little regulatory influence on these events.

Now, we have PCS looming up, and it could impact all of the existing communication services. I believe the regulators and lawmakers will *not* keep up, that the technology will charge faster than they are willing or able to. While this may not cause compete chaos, it is bound to cause some, making investment at both the corporate and consumer levels rather risky.

Enforcement

Where transborder flow of information once consisted largely of shipping printed works, governments were moderately successful at prohibiting the importation of what they did not want. Obviously, the spread of ideas was hard to stop completely, since people with ideas could always find a way to cross borders. But books and newspapers could be banned with some effect. Now information flows in the form of disks, which can be easily hidden, or electromagnetic waves. The waves can be detected, but can be stopped only by jamming a frequency, i.e., broadcasting a lot of noise with a strong signal on the wavelength of the unwanted signals. Countries do not block incoming commercial channels, especially the telephone system, except under virtual wartime conditions.

When the German government tried to ban pornography and the output of a neo-Nazi based in Canada from the Internet, it was easy to get around the ban.[3] The Internet is multiply connected. There is always another route for information to flow through. Generally speaking, it would take warlike measures to stop the flow of information on the Internet into a country.

This does not mean that it is futile to pass laws banning certain types of information. It is against the law everywhere to mug and murder. Yet it is done virtually everywhere. It is not possible to physically prevent the activities. It *is* possible to punish their commission or to eliminate or reduce the activities.

Canada has tried several times to bar entry of certain information, or works containing information, that originated in the United States. In one case it was news of the previously mentioned Canadian murder trial and in another it was a Canadian edition of *Sports Illustrated*.[4] Canadians allege that the content of *SI* is American, produced for that audience, and slightly repackaged in Canada. The Canadian edition of *SI*, then, incurs very little editorial cost and therefore is able to attract advertising away from competing Canadian publications at a very low price. This is not political or moral censorship. It is either protection of a domestic industry or of a domestic culture, depending on how you look at it. The practice of shipping a virtually complete magazine across the border and selling it as a local magazine, with local advertising is easy to detect because, ultimately, the magazine has to be made public, either on newsstands or in the mail. (Canada lost its case with the World Trade Organization.) No one is supposed to know who is reading what on the Internet, so protecting against some particular content is particularly difficult.

The reality, though, is that it is possible for "them" (police and intelligence services) to know who is reading what on the Internet. And "they" may be doing it now. It might be another reason why some people will shy away from the Internet as their principal source of reading material.

The Internet, to repeat, has no means of exerting central control over content. Attempts to use a computer to scan the material, looking for occurrences of certain offensive words, tend to be ludicrous. I have heard of, but must acknowledge I never actually witnessed, one or another form of strange doings when information systems are programmed to ban offensive usages. In one case, the occurrence of the word *breast* in a question to an information retrieval system was prohibited in any context. In others, the words *hymie* and *dame* were proscribed by word processors as being offensive. In these last two cases, they were reported by a person named Hyman who used his own nickname in the signature part of a letter, and another person who referred in a letter to a graduate of the University of Notre Dame. Leaving t to a program to determine what is offensive is not effective.

There is often considerable complaining when someone uses the Internet for advertising. There are two main modes. In one, a commercial service providing access to its web s te inserts ads in its displays. These are just like advertisements in magazines They are there, before your eyes, but you can simply skim right over them if you wish, although they do slow the process of displaying the page in which they are embedded. A second form of advertising is to send the text of an ad by electronic mail. This is much like the "infomercials" we sometimes see on television. It is not always apparent at the outset that it *is* advertising. In e-mail, it fills our computer memories and it takes time for us to detect what it is and delete it. Recipients are often outraged, feeling this is a violation of the eth cs of e-mail and the Internet. But there is no formally adopted code of ethics to prohibit this.

Plain, unabashed ads at least go to support the program or information the ads are associated with. There is ro discernible benefit to the public from e-mailed advertisements. They are at least partially paid for by the recipient's institution. But as with other forms of limit on content, trying to stop this practice has been futile so far.

In the print world it is more tradition than governmental control that has given us some stability, considerable freedom, and yet reasonable restraint on content. Newspapers, in general, still do not print the various four-letter words. If necessary, when quoting, they will venture as far as expressions such as "f—." The *New Yorker* magazine, once rather prudish with text, although occasionally a bit risqué in its cartoons, has now gone whole hog into unrestrained expression. Books can now contain almost anything in print but not necessarily in pictures. Books are not as widely distributed as newspapers and magazines, hence there is some limitation on the audience reached due to distribution cost and availability of the books.

Radio remains close to simon-pure in terms of forbidden words, but definitely not in terms of inflammatory political language or racial or ethnic slurs. (Howard Stern seems to have broken most radio language taboos.) Television has seemingly abandoned any restrictions after the magic hour of 9 pm, when it seems to be assumed that all the children are in their beds. I remember seeing one movie on TV that began at 8 pm and in which naughty words were bleeped out, until 9 pm when the full, colorful language was resumed.

How is this going to end? I suspect a lot of lawyers are going to get some nice fees but that no fundamental changes will be made. Our (North American and Western European) world seems to have become so tolerant of language and explicit depiction of sexual scenes that I do not expect to see censorship of the Internet become widespread. Although this is not a prediction, just a suggestion that this could happen, I *can* imagine institutions that give free Internet access to its members, such as universities and other schools, choosing to impose some restrictions, which I would imagine to be constitutional. Public libraries have long had to walk the thin line between censorship and the exercise of good taste, between serving the needs and legitimate interests of their clientele and the social mores of the community.

Will the Rich Get Richer and the Poor Get Poorer?

It has become a hackneyed tradition in the electronics world to promise that each succeeding invention will make life better for everyone. Among the favorite claims are that the best medical care can be available remotely and electronics in the classroom will cure all sorts of educational ills. I am in no position to evaluate relative states of health or conditions of life among peoples in different parts of the world. I do suspect, though, that in the North American world some of the worst health problems are to be found in the center of our biggest cities, where telecommunications helps less than it must in reaching a remote Inuit village in the far north. I am much closer to education, being directly involved at one level (graduate school), having raised four children, and now seeing grandchildren enter the school system. I do not see anything that communications or computation is really doing to improve the lot of the less fortunate segments of society. But this is a complex subject. Neither blithe claims nor luddite objections are what we need, which is careful consideration of *effects* not of pride of ownership or dreams that machines, not brains, compassion, and taste, are what will do for us in the future.

Equitable Distribution?

Some people believe that access to electronic information systems must become a fundamental right of all people. The telephone is close to this – it has become an important, perhaps indispensable safety device.[5] In some jurisdictions, a family unable to afford a telephone will be government-subsidized in getting the service. Radio is also important in case of major emergencies, such as storms or floods. But radios have become so cheap that virtually anyone can afford one. With the development of the Internet, many believe that access to it should be universal, i.e., available to all, independent of ability to pay. Some progress in that direction is being made by *freenets*,[6] associations that provide free access to electronic mail, bulletin boards, and some other services, at no cost to the user. Equipment is often donated by corporations in the field, workers are often volunteers, and libraries may provide the computers. But as the novelty wears off, the volunteering of money and equipment may wear off, too.

If the world does go in the direction of computers costing a few hundred dollars and communications toward PCS, some people feel we will approach universality. It is still hard for me to see how people who are really poor are going to pay this kind of money for information services. There are just too many people out there, and there are just too many who do not care about information services. If some government were willing to fund the distribution of a computer to every household, we might, indeed, achieve universality. This is not very likely. If computer costs remain in the $2,000+ range, then to the extent that computers contribute to our economic well-being, the gap between rich and poor is bound to increase.

Placing computers in libraries and government offices can help to some extent, but in recent years public and academic library budgets almost everywhere are being cut. Installing, maintaining, and eventually replacing computers is not easy for them. It seems hard to believe that libraries in rich communities will not pull ahead of those in poor communities in offering such services and then, there we are again, with a growing rich—poor differential.

Another aspect of this problem is that if a family is not literate they are unlikely to want to invest in machines that bring in information. By "literate" in this context I do not mean *able* to read but *interested* in reading. If TV came with the package, they would be more likely to make the investment. Does this sound elitist? My own belief is that motivation to read (whatever that word comes to mean), to acquire information and learning, is the ticket out of the ghetto. If families lack this motivation, for whatever reason, more machinery will pull those who are motivated and have money farther ahead. Widening that gap solves no social problems.

I want to avoid making a prediction about which way all this will go. There are too many factors. It is important to avoid oversimplification. I am fairly sure of one thing. A market as big as this potentially is, for a single system encompassing our current computers, telephones, and television, as well as books and magazines, is *not* going to be driven by what is good for the poorer half of society. I rather think the dollar-earning considerations will dominate. I can only hope our government regulators will find some way to avoid disenfranchising or totally alienating those who cannot afford the new technology or whose education until now leaves them unable to appreciate the advantages to be gained by making the financial sacrifices necessary to get on the information bandwagon. In short, it is distressingly easy to visualize a stratified society being made even more so by information technology.

Should Communication Be Free?

Free, here, has two meanings. Free of cost and free of inhibition. I do not hear demands that transmission should be free for all. We do hear demands that the flow of information should be uninhibited. We also hear that no one has the right to shout "Fire" in a crowded building when there is no fire. In other words, almost everyone accepts that there should be *some* restrictions, but there is little argeement on *which*. There are all sorts of culturally based restrictions in various parts of the world.

Is there, or should there be, free movement of information (including cinema and TV as well as books, news, business information) across national borders? Does free movement risk domination of smaller, poorer countries by a few rich countries? Does control over movement violate the precepts of the ever more popular free trade agreements? Can a country protect its own culture by restricting the inflow of foreign information? Should it?

This is another of those questions, or interrelated sets of questions, that is not new, but is made more pressing by the ease with which information can now cross borders. When the German government asked CompuServe, an American company selling its services internationally, to block access to a number of news services that the Germans thought to be pornographic, it triggered the usual reaction. Civil rights groups that do not necessarily favor the pornography objected to any limitations on the free (meaning, here, unconstrained as to content) exchange of information. Others applauded, no doubt including many members of the U.S. Congress who would later impose similar constraints on information provided on the Internet.

Why is this a bigger issue in on-line services than it has been for print publication? At one time, many readers will remember, it was an issue in print, too. To some extent, it still is. Some of the works of Henry Miller and D.H. Lawrence were banned in the United States for many years on the grounds of pornographic content. Gradually, these laws or rulings were relaxed, until now there is very little limitation on what may be published, although there are still limits on what printed material may be shipped across the border. Canada Customs finds itself often in the news over the prohibition of something printed, with heavy emphasis on the banning of homosexual erotic material. Not long ago a Toronto art exhibit was closed down because of the sexual content of some of the graphic art. Canada has hate laws that prohibit the publication of material that contains lies about or encourages hatred toward racial or ethnic groups. The public seems to show relatively little objection to and much support for these hate laws. Any limitation on content based on sexuality seems to remain highly controversial.

Should Information Be Free?

Should all information be free of cost to everyone? This is a point of view often expressed, something of an extension of the idea that the electromagnetic spectrum belongs to everyone. It is a rather utopian concept. If all information were free, then there would be little incentive for publishers to develop their wares. Investing money in preparing a work for publication would go unrewarded and unrecompensed financially. Even nonprofit publishers, such as university presses, must recover the cost of their operations. In our present tight budget world, continuing subsidies from a government or board of directors to cover the entire cost of publishing is hard to envision. Certainly, unsubsidized operation would put more power in the hands of the very rich to produce well-written and -edited material. In recent years both the United States and Canadian governments have heavily cut their support of publishing both of artistic works and science.

Those who favor all-free information tend also to favor unrestricted access to the Internet, *access* in this case used in the sense of opportunity to contribute recorded information. But like it or not, the Internet costs money to build and maintain. Someone or some institution has to pay for it and few are going to take the position that they will pay regardless of what the content may be. Further, unrestricted access means lots of junk. Remember, publishers do not merely carry the money to the bank. They perform other services, prominent among them being to select those works worthy of being published. Of course, different publishers define *worthy* differently. Let us hope that always remains so. But it seems simplistic to expect that all information will be free, or if it were that producers would continue to produce, or that consumers would not be overwhelmed with the volume of useless material.

Gresham's Law and its analogy to information was mentioned in Chapter 4. If information were always free, and there was little consequent motivation to make it better in content, expression, or packaging, would this not encourage a large amount of worthless "information" to be put into circulation? "Free" does not have to refer only to monetary price. There are costs in terms of effort to find information and penalties to one's reputation for publishing debased material. If bad information did drive out good, would not that make it harder for honest folk looking for quality information to find what they want? And, if that were to become the case, would not competent authors or publishers avoid using the information services so debased? I think they would. I don't favor charging for information just for the sport of it, but I do favor investing in the cost of good-quality publishing and finding some way for those who do invest to recover their investments. This does not imply excessive profit, or even any profit. It does imply recovery of costs. Governments, I believe, are morally obliged to provide a great deal of information free to their constituencies. But not everyone can afford to.

Chapter 15
Thinking about Change

When we think about the acceptability or potential of electronic books, it is important to form our judgments based not on today's technology, but tomorrow's. As books evolved from Gutenberg to pocket-size paperbacks, the notion of what a book is has changed. Gutenberg books were not for the average person. They were not for entertainment. They were not even for personal reading. Novels existed but were not common. So why assume evolution in this sense has stopped? We are going to need more evolution before electronic publication becomes the dominant form, and that is very unlikely to happen with today's technology. So, in judging what the future might be like, we have to be willing to assume there will be major technical changes.

Of course, publishers must think about markets – who wants to buy what. What are the forms a "book" could take and who would buy one? We made the transition from text-only to text-plus-pictures a long time ago. Children's books have often led the way. How could some of us have survived parenthood without that multimedia masterpiece, *Pat the Bunny*?

Now we have the options of combining text, pictures (still or motion), and sound. Even the table of contents and index were innovations – technology. Today, we can effectively expand the index to allow us to find any word or combination of words in a text. Such capabilities in a CD-ROM version of Sherlock Holmes enable us to discover that Holmes never did say, "Elementary, my dear Watson."[1] More seriously, we can now not only read the text and search it, but do studies of disputed authorship or of stylistic changes by an author over time. Other modern information retrieval techniques would enable us to find concepts in a large text, even if we cannot think of the exact words to describe them. If authors will write their material in the appropriate way, hypertext software would enable a reader to read in any sequence.

We can combine a history book with on-the-scene video, and with recordings of the speeches of famous people. So, we can not only hear their own words, but search or download the text, see the crowd reaction, and judge for ourselves how a politician's speech sounded, although we cannot project ourselves into the minds of people of an earlier time to sense how they might have felt about a speech.

Are Changes in Book Technology Undesirable?

One of my major contentions is that most of the attributes of current books, as objects, have been with us for so long that we tend to forget that they are merely technological devices; they replaced some earlier form, and they were resisted upon initial development. They are not inherent in nature. The photocopy machine and the verb *to photocopy* have been with us for only about 40 years. Although this may span the working life of most readers of this book, it is not long by comparison with the life of printing, telephony, radio, or television. The point is, there was a world before photocopying. When it came, it affected publishing through the process of producing a manuscript, and there will be a life after this technology is replaced by something else. Similarly for ink-on-paper printing.

But there are aspects of paper printing that are going to change only slowly. There is a lot of emotion invested in books and printing. Most of us who like books as objects think that a good one should have a certain feel – good paper, a good binding, a pleasing dust jacket. Relatively few publishers still include a colophon (a description of the type used or other production data), but it is a pleasant custom for many of us. And the importance of good design, both of type and of the page, remain. Oh, if only more desktop publishers were aware of that! Who would put a junky book in a leather binding, and who would not accord a leather-bound book at least the benefit of the doubt as to worthiness? Just as users come to value the feel of a good fishing rod, pen, or canoe paddle, I think they will come, someday, to cherish the feel of a good electronic book reader. But not today. It may take a generation.

The real issue is not whether electronics will, one day soon, push printing out of the world, but how will one impact the other. We see enough carryover from printing into computing to see the influence there (a screen full of symbols is called a page, for example). How will future developments in electronics affect printing?

Are Changes in Book Technology Inevitable?

In a word, yes. Changes in the relative cost of materials, warehousing, and distribution will make discs or on-line transmission ever more attractive. Some day, we will have an electronic reader that is small, light, comfortable to use, and comfortably priced.

What we will undoubtedly see is a gradual transition, irritating to some as their favorite class of books recedes from the print market, eagerly awaited by others who love working with high-tech equipment. The benefit in use will come first to those who want to look up brief passages ("facts" or quotations), who want or need control over searching or reading sequence, or who want to do linguistic or literary analysis of a text rather than to read it for content. That is why electronic reference works have done so well in the market.

We will almost certainly see a radical change in the methods of storage and distribution of books. One possibility already mentioned: laser printing technology makes it possible (or shortly will) to have a bookstore, school, or even individual user receive text from a network and print it locally. This could eliminate the need for warehousing books and for delays of more than a few minutes in getting one's hands on a desired text. Similarly, any older book could be reconstituted in print, on demand, although someone would have to bear the cost of scanning. The type of binding would be different. Would readers accept this? If so, the role of bookstores and libraries would change, but not necessarily be eliminated. That, however, is another story.

What Might We Want of New Book Technology?

The designs of new media or of new electronic devices for recording on or playing back from media are not based on consumer surveys. Modifications may be so based, but when new ideas are created, it is something like the fashion industry. A designer gets an idea, a company is willing to risk producing it, and with luck the public supports the new product. I set forth below my ideas on what characteristics any new media should have. The characteristics I will discuss are: ease of use, interactivity, versatility, accessibility and connectivity, portability and capacity, cultural continuity, durability, and clarity and resolution.

Ease of Use

There is the horrible expression "user friendly." Why I think it is horrible is that it tends to concentrate only on ease of use and usually on ease of the first use. It does not take into account performance or what the entity under consideration can accomplish. I like to ask people if they pick their friends on the basis of professional competence and if not, why pick their machines, which exist only to do work, on the basis of friendliness.

Certainly, one aspect of this characteristic is the first impression. How easy is it for a novice to be able to use the device or communicate with the medium? The other aspect that has to be considered is how easy it is to extract meaning or actually get work done. Some computer programs are very easy to use, in the sense that all you have to do is answer a few questions in order to get what you want. However, very often such systems cannot accomplish subtle tasks or customize the output. The simplicity of use may detract from the quality, precision, or versatility of output. Painting by the numbers may enable you to produce a recognizable painting quickly, but would not enable you to challenge Rembrandt. Of course, many people don't want to challenge Rembrandt. The point is to consider ease of initial use as well as at the maximum level of intended use.

The first system we learn to use has a tendency to fix our minds on that type system as the way things ought to be. Those who first learned to use a computer with a Macintosh are very unlikely ever to consider DOS easy to use. And actually, the converse may also be true. Even though the

Mac is supposed to be easy to use, for a committed command-language user it is different, therefore potentially harder to adapt to. The fact is, poking at icons is not everyone's cup of tea.

The printed book rates very high on the ease-of-use factor, especially if we separate learning to read from learning to use a book. Learning how to open one, turn the pages, and use a table of contents or index are all rather simple or they are learned at such an early age that most of us cannot remember the learning process. Learning to read is much harder, but we have to learn to read for other reasons than reading books. For an adult to learn to use a computer is often a nightmare, because everything is unfamiliar. It needn't be, but I sometimes believe that even top software companies design for one extreme or another, professional programmers or idiots.

Interactivity

Interaction is something people love. It is better if it is more than just action and reaction. The reaction has to have the appearance of some intelligence or understanding of the action in order to be appreciated. Perhaps that is why e-mail and bulletin boards are so popular on the Internet. You can write something, apparently nearly anything, and you are likely to get some response. When I buy my daily newspaper from a coin-operated box I drop in the coins and then, less than 100 percent of the time, the box is unlocked. That is interaction in a sense: action leads to result. But it is not real *interaction*. If it said "Thank you," or "You are five cents short," I might like it better. If the box won't open, I'd at least like to have it tell me why.

How impressed any person is when interacting with a machine depends on what that person knows the machine is capable of doing. I have designed and evaluated user-computer interfaces, so I am extremely critical of others' designs. But another person might be enchanted by the same system I find inadequate.

In any case, interaction is addictive. Once you experience it you want it all the time. Once you have done some successful information retrieval over a network, and you are then faced with finding the source of a quotation by poring through books on a library's shelves, you tend to be very unhappy. This is one of the great potential benefits of electronic publication. It is not the reading of the text that is the benefit, it is the finding of the appropriate text, and getting help from the system if you have trouble finding it.

Interaction is most helpful when there is uncertainty on the part of the user on how to proceed. This is especially true when trying to find information. Skill at searching for information is not widely held. Often, users of a library do not know themselves what they want, so a good interactive system, whether provided by a computer or a librarian, helps to draw the person out, shows what is available, and describes the attributes provided. It then helps the user rephrase the question based on what was retrieved, and continues in this manner until a satisfactory result is obtained.

I was surprised to see in Bill Gates's book the statement, "The highway's software platform will have to make it almost infallibly easy to find information, *even if users don't know what they're looking for*"[2] (emphasis added). I like to think this was a slip of the electronic pen. If the user doesn't know what the target is, the computer cannot intuit it. If the user knows but has difficulty expressing it, then machines can help. The user has to be a full participating partner in interactivity. Garry Kasparov, the chess master, upon losing a match with IBM's Big Blue chess-playing computer program said that he felt the machine was programmed to play against *him*, not against players in general.[3] In other words, the program might not have been able do as well against any other player of that caliber.

As we have noted, interactivity in fiction and in games can make the reader or perceiver a participant, not just an observer. That is certainly the lure of the electronic games that place the player in a situation and leave it to him to make decisions. (I have a feeling it is almost universally *him* in treasure hunt or war games). There remains the question of whether, in fiction and some instructional material, the reader really gains by varying the sequence of experiencing portions of the work. But I think a good fiction author, in the future, will learn to give the reader a feel for involvement, while retaining control over suspense-building sequence.

Versatility

Future books may use some medium other than paper (or parchment or cloth) and we may collect and store the author's symbols in a manner that permits variation in the sequence of reading. We should not necessarily ignore an author's intent, but if the author chooses to authorize more than one sequence, the book's construction should permit and support this. By "authorize" I mean that the author would so construct the work as to suggest that readers try different sequences and make it easy for them to do so. An example is the use of hypertext, stored in a computer. Another example is a novel with more than one possible outcome, to be selected by, or whose choice is influenced by, the reader. But where it is physically possible to hop around in a conventional novel, the author rarely makes it easy to do so in a meaningful way. Manguel does mention one such book, *Hopscotch*, written by Julio Cortázar, in which chapters were written to be selected in any order by the reader.[4]

If the recording medium is a computer, or a component of one, or some similar device, then we can have some additional features. A traditional book can be annotated by its reader. The most exciting bibliographic moment of my life was seeing and touching a book with marginal notes written by Isaac Newton. To the annotator, annotations add value-in-use to a book. To the collector, annotations by Newton do fabulous things to the market value. But to the reader who did not make the notes, and assuming the notes are of no historical importance, annotations range from valuable commentary to amusements, to irritations, to vandalism. Could we not have a book wherein annotations are freely permitted, but remain private to the annotator, or at least optional to the reader? You can do something like this, today, with a word processor, but it would be awkward since you would have to maintain a separate file for the notes. I'd like to be able to write on the text I'm reading and then decide if I want these notes saved for me only, or for the public, or any sub-public I care to define and any reader could elect to see or ignore them. In an age when faxed representation of a signature is accepted, we may find that original, handwritten documents may disappear from the market – the facsimile is as good as the original. Gresham's Law again.

Tables of contents and indexes, as noted earlier, are technological devices invented after the book itself. To an average modern-day reader they are immensely valuable. We can imagine what the pre-index reader had to cope with by considering the average computer manual, whose indexes tend to be examples of the concept *useless* because they omit so much. But these artifacts are limited. Could we search the text for occurrences of any word or sequence in which we are interested? Would that be useful? Could we search for any *topic*, even if we do not have the exact words used in the text being searched? Again, a common problem of computer books and computer-stored help systems is that the reader must know the author's terminology for some concept, not just the reader's own terminology. If you want to know how to store a file you have to know whether to look under *store* or *save* in most systems. They rarely offer both options. When we read a passage, whether in a novel or text book, that contains words or symbols we do not understand (or have forgotten), how nice it would be to call up the definition or previous mentions with a click or two of a button or mouse. How nice, also, if the definitions presented were in the context of the book being read, so *field* in a physics book would be defined differently from *field* in a book on athletics or agriculture.

In technical books it would be desirable to be able to ask the question, "How do I do ...?" A likely response to such a question is probably not a prepared paragraph of text, but a sequence of paragraphs or other text sections selected in response to the specific question. Can this be done now? We are seeing such methods installed in the help facilities of some software packages. My experience is that this is basic information retrieval, not interpretation of what the searcher means. The system looks for occurrences of the words you use, it normally does not attempt to find other words or to deterime what you *meant* rather than what you *said*.

Yet another aspect of versatility is related to custom publishing (see Chapter 10). Much of the same mechanics of assembly are used if a student is compiling a term paper by extraction from a variety of sources, rather than by writing the material himself or herself. As the volume of material readily available on the Internet grows, it becomes less and less likely that a teacher will be able to recognize the sources used, hence to detect this manner of "writing." It might be possible to embed a software seal of some sort into a text to assure that if ever displayed, the source note is also displayed, as proposed by Nelson.[5]

Accessibility and Connectivity

Electronic publishing can give us not only quick access to segments within a work (as in hypertext) but quick access to other works, wherever they may be stored. This is another of those "benefits" whose value is hard to estimate because we have been raised without it. I recently tried to buy a book from the U.S. Government Printing Office, only to be informed, after several weeks, that it was out of stock but was available through the government's depository libraries, none of which are in Canada or near me in the U.S. If it were on the web I could "get" there in seconds. This is too appealing a prospect not to want to add it to our wish list.

We must remember, though, that what is on the web is *not* all that is in the Library of Congress or the British Library. Their catalogues may be available, but not the content of the books, not yet. As we work toward the day when all the world's recorded literature *is* on the web or some subsequent equivalent, we are going to have to learn a whole new way of seeking and working with information. Imagine writing an essay or term paper in which you are held responsible for knowing what anyone in the world has said on the subject! Previously, this would have been, implicitly, limited to those whose words on the subject could reasonably be found in the library at hand. Today, although the information available is changing, it may well be that the phrasing of academic assignments is not and implicit limits will gradually be broadened to encompass the world.

Portability and Capacity

One of the oldest verses many of us ever learn goes, in part,

A book of Verses underneath the Bough
A Jug of Wine, a Loaf of Bread – and Thou
Beside me singing in the Wilderness –
Oh, Wilderness were Paradise enow!"[6]

Omar Khayyam, the author of these words, was clearly not envisioning a Gutenberg book, King Henry's Great Roll of the Pipe, or a Hebrew Torah, but expressing a clear interest in portability of books of verse. The image is not too meaningful unless we can imagine his book to have been something of the order of today's paperback or the small-size leather-bound books popular in the late 19th and early 20th centuries.

What gets read on a subway? Mainly newspapers, especially tabloids, and paperback books. Oh yes, I've seen people working over spread sheets, but most of us don't get enough space for that. Even full-size papers are usually folded. (One of my manuscript readers inserted a note in the margin, here, saying, "In our city riders reload their guns.")

I am sure that one of the reasons for the resurgence of popularity of radio, after television became well established, was its increasing portability. There must be hardly an automobile today without one. The same goes for the audiocassette. It's no longer a teenage phenomenon to walk down the street listening to a radio or cassette, using a Walkman. The "suits" do it, too.

Motion pictures have kept up their popularity reasonably well in the face of competition, but they are hardly portable. Producers and exhibitors have to offer something else. So they gave us bigger screens and better sound systems. It does matter. The wraparound screen gives a different sensation from the old-fashioned one using only the center of a stage. The higher fidelity and stereo effects that make sounds come from different directions give a greater sense of realism – this is not just for music. Modern theaters also give us popcorn machines and a social environment. A comedy, for example, is more fun when there is a house full of people laughing together.

Computers are called portable, but the laptop weighs in like a large dictionary and the pocket-size ones don't really do very much for us, yet. So, on this count, too, the book seems comfortably ahead of its competition.

Like many people, I read while seated at a desk, at the dining table, lounging on a sofa, in bed, in an airplane, on the beach, etc. Many of today's books are beautifully well suited to such use. But not all, and certainly they have not been so all through their history. Thus, convenience has not always been part of the definition of a book. To repeat an earlier question, who would care to read a Gutenberg bible in bed? No electronic version of a book that I know of today has the portability of a typical paperback book. But, as just pointed out, neither did early books. SONY Corporation's Data Discman, brought out in the early 1990s but rarely seen in North American stores, approached the paperback in portability, but its viewing screen was only 65 by 56 mm (roughly 2 ½ by 2 ¼ in.) and I would never try to convince a modern book lover that this is to be preferred over a good paper edition. But give it time. Paper book publishers have had 500 years in which to improve their products. Given such handheld computer systems as Apple's Newton, can we not envision a truly portable, convenient, electronic reader/display in, say, five more years?

This has been mostly about the computers used to read materials. The recordings, themselves, are going onto ever more compact media, and there is every reason to expect this trend to continue. SONY used a compact disc with its Data Discman that could hold over 100 complete novels. A full-size CD-ROM can hold even more, a DVD a good-size library. Even so, these are not quite to the level of the Library of Congress but they can easily handle all the readings assigned in a four-year university program.

Cultural Continuity

I don't know that many people would consciously list this as an important aspect of media or something in which to guard against a change. To a significant extent, a culture is defined by its literature, which is not limited to the written word, but can include oral traditions, music, and graphic art. The more media we have and the more variant forms of literature we produce, the less do we, members of a community or culture, share the experience of the same songs, stories poems, pictures, and novels.

I don't think we can blame the producers of electronic publications for damaging our culture. They have not set out to do this, but I cannot help thinking that their products may have this effect. Perhaps, this would even be a good thing. We've certainly had enough wars and killings caused by ethnic hatred. Maybe lessening the differences by lessening the concentration on certain shared experiences will ease some of the tensions. But that's beginning to stray too far from the focus of this book.

A relatively small point that still bothers some people is that books are usually produced by creating a manuscript, editing or modifying it, typesetting it, then correcting the proofs before the book is finally ready. Subsequent editions may show further change. Scholars and rare book collectors treasure the record of these successive metamorphoses. They may all disappear with electronic publishing. While word processors can make copies of every change, no one I know keeps all the different versions. It may become hard to tell what is the latest version of a text or how the author progressed from an original draft to the finished text.

Durability

"Guaranteed for life." ' Guaranteed for the life of your car." These sound like long times, but we expect much more from books and artwork. The *Dead Sea Scrolls* have lasted a couple of millennia, even if they have to be treated now with extreme care. Gutenberg books survived half a millennium and anyone who owns one would treat it with extreme care, but it is also true that these books were rather sturdy and might be able to tolerate some rough handling. A book that is a mere hundred years old today gains almost nothing in value from that fact alone.

On the other hand, newspapers begin to crumble in just a few years and many books, made from acid-containing paper do the same. We began to realize in the 1960s that we could not just put magnetic tape on the shelf indefinitely, that it would deteriorate. We have had experiences of microfilm interacting with chemicals in the file cabinets and deteriorating as a result. CD-ROMs *seem* to be long-lasting little things, but we will have to wait a long time to be sure. One danger is that, even if the disc lasts a long time, there may be no reading devices in, say, a century from now, because the technology changes. If you had an old Edison cylinder sound recording, the precursor of the phonographic disc, you would probably not have a "reader" or playback device, and you certainly couldn't play it on your CD player. Yes, you could go to a museum to find an Edison player or, someday, to an old electronic bookshop to play back your old discs, but this is an age of people wanting things at their own fingertips.

What do we want in the way of archival durability of recording media? Most of us do not really know, but probably have some sense that someone or some institution ought to have an archival copy of all important publications or records. And it would be nice if our own copies could be maintained more or less indefinitely if we chose. I would like to do better than current floppy discs. CD-ROMs *seem* to be good enough.

Clarity and Resolution

One of the great advantages of printed books over almost all computer systems is the clarity and precision of reproduction, whether of type or of graphic images. A book page with well-selected type, sharply printed on good paper, with appropriate margins is much easier to read than the display on a typical computer screen or even the output of most computer printers. Computer printers are catching up, but we are not far removed from the day when computer output meant printing with a dot matrix printer and no choice of fonts. The motion picture screen provides a similar advantage over the television screen, and FM stereo radio provides a similar advantage over AM radio and telephone in sound reproduction.

Much of the desired resolution could be provided today for computer screens, but the cost would be prohibitive. Similarly, it is possible to have a newspaper-size screen, but if in a fixed position, it can be hard to read, while we can fold a newspaper any way we want, to read a single article at a time, but scan a full page if that is what we want. I have seen mention of just such a foldable plastic sheet as the computer output device. Coupled with a highly compact disc reader this could make a big difference.

Until about now computer output has not really competed with anything. People were glad to get it, in any form. Now that the computer world wants to compete with the print world, it is going to have to be able to provide the clarity of image that cold type printers have been giving us for hundreds of years.

What Do We Want?

Early in my career as a systems analyst, I learned the risk or even folly of asking people who do not know much about information systems what they would want their new system to be like. The reason is the difficulty of people who really do not understand how a technology works imagining what a new system would or should be like. Even harder is to foresee how they might adapt to it over time. Major progress in information systems is usually brought about by a technical person or team who get a good idea that catches on, first with financial backers, and then with users. There are, of course, lots of ideas that don't catch on, but most of us remain forever ignorant of them. Among those that many people have heard of but which did not last are Betamax video recording and 8-track audiocassettes. Recently, Apple Computer has been in serious financial straits and could disappear, along with a technology that their many users consider far superior to that of the IBM-Microsoft version of the PC. What we actually see tends to be a steady stream of winners.

When asked what they would like to see in future systems, users or potential users tend to name minor variations on what we already have. What we accept is a combination of what appeals, what is heavily promoted, and what has caught on so well with others that there are few alternatives.

Both spreadsheets and word processors caught on early in the personal computer age, imperfect as they then were. We cannot say they filled a need, because the specific need wasn't felt. But they did provide a capability that many people immediately recognized as valuable. Thus, appeal is what sold them.

I think of many Internet features as examples of heavy promotion. Not all of the hype is commercial; much is done by users not only satisfied but obsessed. This does not mean that the Internet is not a "good" system, but that peoples' belief in it sometimes seems to exceed the reality of what it provides today. What we should not be doing is setting up unreal expectations for computer users. People will still have to know how to read for comprehension, compose text, and conduct searches. Programs do not do these for us, they do them under our guidance.

Windows, the Microsoft operating system for IBM-type personal computers, and, even more, Windows 95, its later version, became immensely popular, and there is still much debate over whether this was entirely due to merit. What is fairly clear, though, is that any PC owner who elects to go with any other operating system is taking a risk of not finding adequate support or an adequate range of software products that can be used with any other operating system.

Preparing to cope with change before we even know what we are changing *to* is a risky business. Young people today are more used to rapid change, and I suppose it has always been a characteristic of our species that younger members adapt more readily than older members. Older members will have seen more disasters and are less ready to hop on the next unproven bandwagon to come by.

Chapter 16
Thinking about the Future

Predicting the future is, in many ways, a waste of time unless you make your living betting on sports or the stock market. There is no penalty for just writing about it and being wrong and, often, not even a payoff for being right. There *is* a payoff for predicting a glamorous or shocking future. It sells more books.

If prediction is to be attempted, it is essential for both prognosticator and reader to coordinate their visions of future technology with those of the nature of the people who will use it and with the time periods involved. For example, one of the most common reactions to the prospect of electronic publication is horror at the thought of having to read favorite books with a current computer. No one I know of expects that to happen. People who predict computer-based books, whether or not they favor them, think in terms of new kinds of computers, but don't always take the trouble to say so. Hence, proposer and listener often have different models of electronic publications.

I see four main aspects of future development to consider: the *technology*; the *publications and the people who will use them*; the *effect on society* of changes in the first two; and the question of the *future of print* itself. Actually, this last question is related to the previous ones but it is so important and emotional, it is considered separately in Chapter 17.

Technology

Technological predictions, while easy to make, are often wrong and are perhaps best remembered when dramatically wrong. Digital Equipment Corporation had a television ad in 1995 that stated, "There are many visions of the future. Most will turn out not to be right." Why is this? After all, many of the predictors are smart people. I think there are three reasons. First is the decision on what technology becomes successful, i.e., is adopted by many users at prices that keep the producers in business. Second, success depends on buyer preference, not proven product superiority. Third, new information, communication, and transportation systems tend to change the environments in which they operate. Hence, the technology is introduced into a society, the society changes, and then all bets are off on how the old society would have reacted. The old society isn't there anymore. The technologies do not just change their direct users, but whole communities of those who interact with the users.

The literature of the history of science and technology abounds with what came to seem ludicrous *after* some particular change came about. There have been declarations that the telephone, the Atlantic telegraph cable, powered flight, running a four-minute mile, and flight by bumble bees all were physically impossible.

Thomas Watson, first chairman of IBM, was said to have predicted only a minuscule market for electronic computers. In the 1950s and 60s, IBM was nonetheless spectacularly successful in making and selling them. Many observers felt IBM's products were not necessarily the best but their marketing and customer support were the best, without doubt. Further, customers could be sure IBM would be in business next year while, in those days, you could not say the same for many of its competitors.

The uses of the telephone, when it was first invented, were not easy to see. Arthur C. Clarke quotes the then engineer-in-chief of the British Post Office as denying it had any practical value. "I have one in my office, but more for show ... if I want to send a message ... I employ a boy to take it."[1]

As noted earlier, 8-track audiocassettes never made it big. Their smaller replacements did. Betamax did not last as a popular form of television

recording. Its competitor, VHS, did. Videophone was first offered by
AT&T in the early 1970s, but never caught on. Cellular telephone was an
almost instant hit in the 1980s and PCS may top it in the 1990s or 2000s.

The development of the telephone, elevator, electric train, and automobile changed the whole pattern of how we organize our cities. Word
processors have changed the way many of us write and edit and the way
we produce almost all printed works. Hypertext may change the way we
read. Computers have changed the way we bank and buy airplane tickets,
and may yet radically change the way most retail sales are conducted.
The behavioral consequences of these latter-day changes are still unpredictable. For example, a major switch in retail buying from in-store to
catalog shopping via the Internet would affect store employment, shopping center economics, the distribution of advertising revenues, personal
and mass transportation patterns, and, as a result of the transportation
changes, the whole pattern of neighborhood development.

I will not attempt specific technology predictions. Based on observations and predictions of others, it does seem that the following are likely,
hence rather safe predictions, partly because some of the technology is
here now:

Computer memory and speed will continue to increase. Moore's and
Grosch's Laws[2] suggest these will double about every two years. As
a practical matter, this means that system developers needing more
speed or storage need wait only a bit for these to be routinely
available. For users, it means that their computers will continue to
become obsolete every few years, but the new ones will do things
that hardly seemed possible a few years ago. Unfortunately, the
computer hardware and software industries depend on this rapid
obsolescence as well as on expanding markets. It is to their advantage
that the situation continue.

Display technology development will accelerate. It has not really
advanced as rapidly as have computer memory and logic, but it will,
eventually, produce a fairly flat screen, approximately book size,
and with resolution fully as good as that of a well-printed book. Book
pages, of course, vary in size, and so will computer displays, from
something easily portable to something hanging on a wall. Probably,

that wall hanging will also serve as the family's TV display. Probably also, it will not really be flat for long, but will project three dimensional holographic images, perhaps through a user's spectacles.

Software technology is impossible to predict. We really do not need anything new to be able to read an electronic book, and we have good search software to find that elusive quotation or person's name in a large volume. But other kinds of programs will develop, certainly providing improved handwriting and voice recognition. I am skeptical about computer reasoning power compared to what we see in fiction, but not about a machine's eventual ability to recognize which word was spoken or written and to suggest possible subject connections of one document to another without explicit links.

Telecommunication will progress as fast as computer components. This means, among other things, that fast, economical transmission of a book-length manuscript or full-length motion picture will pose little challenge. Will our progeny's ability to use information grow as fast as their ability to transmit it? That remains to be seen. The terrestrial distance between origin and destination will become of no account in reckoning cost.

Publications and Their Users

By *publication* I mean works of the human mind that are produced and distributed in some manner, regardless of the media on which recorded or through which transmitted or displayed.

What will be the effect of the changes in publication brought on by computers and modern communications? I will suggest some possible directions of change, what *could* happen, rather than what *will* happen, with emphasis on *effect*, not technical development. What is really important is to try to understand the relationships among the technologies and the various populations affected by them.

The transition to electronic publishing will probably be most helpful to readers in nonfiction. But don't forget that the people change along with the technology. Future generations (I don't see all this as something that would happen in a span of just a few years) might find the change to electronic, multimedia fiction fully acceptable, even preferable. That is often a point of misunderstanding between those who favor publishing going electronic and those who hate the idea. The change won't happen to *us* in *our* time. It will happen to some *other* people, in *their* time.

Effects on the Creators, Publishers, and Distributors of Knowledge

McLuhan decried linear writing and its effect on linearizing and rationalizing thinking, where rationalizing was not always considered a positive act. What does the capability to produce literature for multiple senses and in nonlinear form mean to us? How will authors adapt? How will their works be distributed? Directly to users? Through retail stores? Through libraries? How will these institutions be used?

McLuhan, Innis, Logan, and others pointed out that Western society took a different turn after "we" invented the alphabet.[3] It changed our life style and we became a logical, scientific, linear-oriented people. Will something like this happen again, due to electronic "printing?" It could lead us *away* from linear thinking. What would be the effect of that? The alphabet-induced change in behavior if, indeed, it happened that way, took a very long time. Things change much faster now. Will any change be perceptible in a single lifetime? Will the change be beneficial? Will our successors on Earth have a different idea of *beneficial* than we have? It is quite safe to make some predictions, because I do not expect to be here to be called to account if I am wrong. Our progeny will not *have to adapt*. They will grow up in a world in which the new forms are already dominant – nothing left to adapt to. Will it be beneficial to them? I think they will think so, because that's the way they will be.

The Creators of Knowledge

What ever happened to orators and original storytellers? Their skills are not in high demand in this day of so much televised and edited communication. Might the same fate befall print writers? Will our best print novelists and poets simply convert and become multimedia artists? Not likely all at once. Might they become, like painters, respected artists but whose works are seen and appreciated only by a small elite?

We might well see the Negroponte-predicted transition from the several different media of today into a single medium, which we now call multimedia. Authors would have to make the transition also. They would have much to learn to achieve the level of artistry in the several media of today's specialized creators. It will take time and will probably involve the next generation of authors, not the current ones. This is a major reason why the transition from many media to a single medium will not happen over night. Perhaps we have seen the beginning of this shift in the numbers of novelists who become screen writers.

The Publishers

In distribution we have almost a paradox. One possibility, the one most favored by the technological prognosticators, is that distribution of publications will become so cheap they will be almost free, and everyone will have access to the publication process. In this model, distribution would come to mean little more than making the multimedia equivalent of a manuscript available on a network for downloading by customers. We could then expect a great many more published works than we have now, when almost everything has to go through some kind of filtering or selection process and requires a monetary investment. Not just the computer people, but the telecommunications people lick their corporate chops over this view. Publishers do not.

Another possibility is that preparation of the work, the creative processes of authoring and publishing, will become increasingly difficult as more and more media are involved. They will have to include print, sound, motion pictures, and possibly virtual reality, all to be carefully coordinated. The preparation of a publication will become more difficult and expensive. It is only distribution of the end product that gets cheaper. The higher production costs would limit the number of producers who require a monetary investment and a return on their investments. The electronic equivalent of the vanity press seems to thrive on the uninhibited web. Perhaps the need for multiple skills in future publication is one of the motivations for the formation of conglomerates of media companies that we are now seeing. If the market is dominated by an ever-shrinking oligopoly, why should we expect a greater richness of product?

Ron Howard, the cinema director, pointed out that at one time the decision makers in the motion picture studios were the owners. They could make mistakes and lose money but not be fired. They were willing to experiment. Today, decision makers are employees. They *can* be fired. They tend not to take chances with new ideas.[4] Books or multimedia publications could go the same way — that is, stagnate.

Remember that publication involves more than finishing the product and announcing it as available. The producer has to have some way of getting that announcement to the world. It is naïve to believe that the giant conglomerates will welcome freelance, free-of-charge contributions along with their own productions. Even if they did, there would still be the problem of letting the world know what is there. Advertising is one method, but that costs money and here we are talking about free or virtually free publications, or at any rate publications without a publisher willing to advance the cost of production in anticipation of a profit or cost recovery. Searching the networks by users is another way for them to find out what is available. We'll discuss this later in the chapter, under *New Skills Needed*, but remember, "they" keep telling us how much there is going to be out there, so finding unadvertised or unreviewed works that are just what we want may become quite difficult.

There may be other effects on publication. One is that a publishing company's employee skill mix is likely to change as custom publishing and perhaps other forms of cooperative publishing continue to emerge. It will be important to have skill in finding what's already available that can be sold in a different package, and to find ways to repackage material, rather than to look only for completely new works. The mechanics are also different, since these custom published works are typically not put into the same kind of package as a conventional book.

Will there be more, or less, competition in the publishing industry? The old-line companies show a great affinity for mergers. But new companies always seem to be popping up. There is less and less capital outlay required to become a print publisher as printing and distribution processes become ever more electronic. Negroponte says, correctly, that telecommunication of bits costs less than transportation of atoms. My guess is that we will see something like equilibrium. Existing companies will be consumed by others at about the same rate that new ones are born. Perhaps the birth rate will be higher than the consumption rate because some will die a natural death caused by lack of revenue. But dominance by a few giants, short of total control, seems the clear trend.

What about prices? Manufacturing and distribution costs can and should come down with automation. Paper prices have been volatile recently but have backed away from their steep rise of a few years ago. If the electronic products can be made attractive enough to consumers, the cost per work should come down. However, producers in other fields tend to make up for lower basic prices by offering more options and extras and consumers seem to cheerfully adapt to having and paying for the increased offerings. Perhaps then, instead of less expensive books and magazines, we may see books and magazines with included videos, stereophonic sound, maybe even virtual reality, and these will bring the prices right back up to what we are used to.

Variety is one of the determinants of price. More options, as just noted, cost more. Consumers want more options, or at any rate they respond favorably when offered more. Think how the numbers of models and brands of automobiles have grown in recent years, even though the number of manufacturers is virtually constant. Think of the shift of department stores to being more like a collection of boutiques, seemingly competing among themselves under a single store's umbrella, and of the variety of electrical and electronic products on the market. Centralization might be a counter to this trend – why offer more if you have a virtual monopoly? But I don't think we'll actually reach the monopoly point in publishing, where a single firm controls the industry. As long as there is some competition, offering more options than the competition is always a good strategy.

The Libraries

Libraries seem to have been fairly successful at convincing legislatures (at all levels) to provide funds for computers and networks, but not so much for continued collection building and staffing. Electronic distribution and remote database searching can change the economics of libraries and to some extent already have. This kind of change, however, is going to exert a great deal of pressure on publishers. One direction of change as libraries become more mechanized will have to be an increasing demand for electronic publications, assuming these would help offset the cost of library automation. Electronic publication, where the user gets access only to a copy of the work, never the original, can also be a form of preservation. Theft and vandalism are increasing problems for any print-based information service. Natural disasters, such as fire and flood, continue to occur. On the other hand, publishers seeking to protect their investment in print publications will not switch to a cheaper form without some means of protecting their revenue. We discussed one aspect of this, electronic forms of scholarly journals, in Chapter 10.

Libraries will be faced with ever more works in their collections.[5] This means more pressure on reference librarians to help people find information and to access other collections. They also must get the equipment needed to aid users in browsing and reading. I am not the first to have suggested that the granite library building will gradually become obsolete and that reference services, perhaps provided via telecommunications, will take over as the principal service of a library. The electronic files can be housed in an inexpensive warehouse or subbasement. Only a few employees would need ready access to the files.

We are already seeing the beginning of every library becoming a repository for all the world's published information. Not, of course, that all printed books and journals would be physically present in every library, but as this trend continues every library would have access to every other one's catalog and to a huge amount of material on the web and other online services. This could have several effects. One is in cost, one is in the physical size of the local collection, and one is in the nature of the service that each library would be expected to offer.

It is tempting to say that the cost of operating a library would come down because electronic distribution is cheaper than that of paper. It remains to be seen if the prices will in fact come down. They have for the electronic encyclopedias, so it is a reasonable possibility. But the electronic hardware is expensive to buy, maintain, and periodically replace.

What would we users expect of an electronic library having access to *everything*? I would expect reference service par excellence. I would want a reference librarian just as familiar with what is published in Italy as in Canada or the United States. Or I would want ready access to libraries in Italy, and some help in overcoming the language problem. This would place a great new burden on reference services, and is the reason I do not think libraries will disappear, just their paper collections. Oh, how nice it would have been for the fictional gentleman portrayed in IBM's advertisement, described in Chapter 10, if the entire library had actually been digitized.

The Bookstores

I think of bookstores and libraries, today, as rather similar services. You go to one to buy works, the other to borrow or consult them. In both, you expect intelligent advice and interpretation. In both you expect help in getting a work if the local facility does not have it, which is called either special ordering or interlibrary loan.

If electronic publishing comes to mean publishing on discs, there may not be a big change in bookstores. Currently, many owners, especially of smaller stores, are worried that disc sales would go largely to electronic discount shops, where they are most commonly found today. But these hardware-oriented shops do not tend to offer advice and interpretation (e.g., What's good reading for me or Aunt Millie? What is the best encyclopedia for my children?). That advice is valuable, and continuing to offer it could save the bookstores.

Today, trade books are usually shipped to bookstores on consignment or under some terms such that the store may return unsold merchandise for a full refund. The publisher suggests a retail price and charges the bookstore a portion of that, usually about 60 percent. The shop is free to charge any price to the retail buyer. Discs, on the other hand, are sold as hardware; the retailer pays the distributor a fixed price, which is about the recommended retail price less 25 percent. The actual sale price is usually below the nominal price, making the retailer's margin even less. At first there was no return privilege. Now, the terms are beginning to change.

Bookstores selling large numbers of electronic publications would also need computers on hand to enable customers to browse, and they would have substantial staff retraining costs. The retailers must take a bigger financial risk than they are used to. Can the trade survive so much change? These questions are a large part of the reason bookstores have not picked up these new products more avidly.

For publishers, the right-of-return tradition is a costly one. They would like to see the end of it, but that might also be the end of bookstores. It is essential for all of us who like books that publishers be able to make a profit that will keep them going. In truth, there are other ways to market books than through bookstores, but they must be published to exist. There is an interesting parallel with recent news that airlines are cutting commissions to travel agents and are beginning to use paperless ticketing, thereby reducing both the need for travel agents and the latter's income. The travel agent performs a role quite similar to that of booksellers – retailing a product or service produced by someone else, and offering their clients consultation on what to buy. Will we continue to use travel agents if we have to pay for the consultation? Or use bookstores if we have to pay for their service? Or are both airlines and publishers shooting themselves in the foot by doing away with these valuable retailers? It's still an open question.

If electronic publishing turns out to be mostly on-line, or uses the Internet or some successor network configuration for distribution, then bookstores might, in fact, be doomed. There would, however, still be a need for helping users find the appropriate work, but I doubt we would find as many publication-advice-for-a-fee shops in existence as we now have bookstores. This might be a service that ends up in the public library. When I say not many, I do not mean none. There are services today that find information on behalf of clients, for a fee. The clients are usually corporations, research institutes, or law firms. These services could expand their scope to service individuals.

Effect on Education and Informal Learning

There are many people willing to assert that computers in general, and multimedia-capable, networked computers in particular, will revolutionize education. What they can do for us is allow more individuality in instruction, especially when instructors give students a broad goal and let the students find the information they need and construct from it their own models of the subject matter assigned. By participating in communication networks they can seek out resources in distant places, interview people, work in teams without the need to meet physically in one place,[6] and cooperatively build the model. Students can be freed from the stultifying effects of lockstep education in which everyone does the same work at the same rate, frustrating alike the quick and the slow students.

In Chapter 11, I pointed out some possible negative effects, as well, mainly that the least imaginative students might not gain as much from the freedom of action or that some might substitute extraction for imagination.

By no means do I suggest that a shift to all multimedia, self-directed education would *necessarily* lead to a world of only unimaginative people. I say it *could*. For many, multimedia educational materials could represent a gain. We might gain in average educational accomplishment. For those with mental or perceptual handicaps, it might be a boon, because multimedia works can be more versatile, more readily adapted to the individual's particular needs. It's just, well ... be careful. A change in the wrong direction could be hard to reverse.

Schools will have to change some methods of teaching. They will be offered "marvelous" new educational materials. They will have to invest ever more money in educational technology and if they don't have it, will fall behind in the esteem of parents and legislators. They could still teach reading, writing, arithmetic, and the arts, but these will lack glamour and many parents would complain about that, too.

Reading

I believe that reading in this new multimedia world is or will become a different activity than it is in the print world. I believe we will need new skills and different methods of teaching reading. What would the transition do to readers? Will our children lose or never acquire skill in and love for reading the printed word? Reading print has been important to educated people for hundreds of years. (Important note – not thousands; it was an acquired taste.) It requires a great act of faith to believe it will all come out OK. But, however it comes out will be OK with the people who live with "it." There are those today who cherish the oral tradition and feats of memory of the ancients and of native North Americans right up to fairly modern times. But most of us cherish reading more than feats of memorizing or storytelling. Our grandchildren or great-grandchildren may cherish something else.

The positive side is that, at least initially, electronic reading is intriguing and engaging. Many people simply enjoy browsing the World Wide Web or getting involved in hypertext. These activities might encourage those not otherwise inclined toward reading to find interesting reading material and become inclined.

New Skills Needed

To function competently in the new, high-tech world, the average user will need skills not universal today. This applies to practicing professionals as well as students. We need to know how to express ourselves, in print, orally, and, to an increasing degree, graphically. We need to know how to read, not just to recognize words but to *understand* what was written. We need to know how to listen, again not just to hear, but to understand. We must not let ourselves ignore the intellectual aspects of these activities in favor of the mechanical aspects.

I see an eventual need for the person considered educated to have the characteristics defined below:

literate. 1. Able to read and comprehend written material and to competently express thoughts in written form. **2.** Having a good knowledge of the best of printed works.

numerate. The ability to read and understand materials that present data quantitatively and to be able to express data quantitatively.

computer literate. Essentially, able to use a computer, or plan for its use, productively. A term now in common use, coined by Drucker.[7]

graphicate. (Not a word in your dictionary.) The ability to comprehend and appreciate information presented in graphic form and to present ideas in this form. Graphic form means not just graphs of quantitative data, but "art," as well.

information literate. Knowing how to deal with information, particularly how to evaluate it and to be adept at searching for it.

Put them all together and they make a person able to work in all the modern media. That does not necessarily mean being an artist. It could mean being a lawyer. Lawyers have always had to be able to read comprehendingly and write both understandably (at least to other lawyers) and convincingly. Now or in the future, they will have to become adept at using graphics in the presentation of a case to a jury, to deal with probabilities that forensic scientists drop on them, and they have always had to be able to find information about precedents for a case.

We need to know how to search for information. As incredible amounts of data are made available to almost everyone, almost everyone is going to be expected to know how to find what is needed. Excuses like, "I've never seen that before," once said to me by two physicians in a practice that had misdiagnosed a major problem in a family member, or, "It's not in the library (telephone directory, today's newspaper, etc.)," which I hear often enough, will not be long tolerated, because if you can't do the job there will be someone else around who can. In our culture, searching is not taught as a subject unto itself, except in certain graduate school programs (law, library science, chemistry). We others are given superficial lectures on a library's classification system, but rarely taught how to search for some fact or to find information that is not in a library. The Gates quote in Chapter 15 about how easy it will be to find information even if we don't know what we are looking for cannot hold up. The machine does *not* do it for you. It cannot know what you had in mind, only what you said, and these are usually different. You have to know how to direct the machine to find what you really want. It can help, but not do the job for you. On the other hand, finding ways to help people to search for and evaluate information will be a major focus of computer research in the coming years.

None of these skills is specific to computers or a high-tech culture. They are useful at any technological level, anywhere. As with library service, the growth of information systems will place greater, not lesser demands on people to be able to find and comprehend information. Possibly, at some far distant time, computers will really be as smart as people. Then they will have to become as understanding as people. I don't have to worry about seeing that day. If you are old enough to read this book, neither do you. It may be another legacy for our great-grandchildren.

Lifelong Learning

What started as something of a cliché and, as I always suspected, an advertising slogan for colleges and universities strapped for money, "lifelong learning" has become an ever more realistic, appropriate slogan. With so much changing around us, we have to struggle to stay in place. For mature, motivated learners, computer-assisted instruction, distance education, and other forms of instruction through the medium of computers hold much promise. We will not have to enroll in and attend a school, demonstrating our ignorance and insecurity to younger people. We can take courses according to our own schedules, receive text materials, and even do simulated laboratory work through the web or VR or just plain computer multimedia discs. Whether computer-based or not, the future will belong to those who have the initiative to seek out continued learning.

Just as the price of paper could accelerate the push toward electronic publishing, the lowering of school budgets and increasing of pressure on teachers to do more with less help could accelerate the use of electronic teaching materials, as administrators cling to the belief that they will cost less. I am less sanguine about this for small children than for adults. In fact, I see no great need to introduce small children to computers. They need the human contact and they need to learn the fundamentals of reading, writing, arithmetic, art, and science. Adults can deal better with the impersonality of computer learning. Computers can be used productively to drill students and analyze their performance, perhaps better and faster than can their human instructors.

Distance Working

Might we reach the point where most of us stay home and do all of our business and entertainment remotely? It could be, but I think that people are not ready for that yet, at least partly because they don't yet know how to use all that equipment to its fullest extent.

What if we reach the point where most people *do* know how to use the electronic media? I think any of us who like human contact are uncomfortable with the prospect of dealing with others almost exclusively through the medium of a machine. But there are lots of people who are more than comfortable, they are excited about it.[8] One aspect of this working at home question is that communication via e-mail or even television is not the same as communication in person. We miss the subtle gestures, the chance to go for a walk in the woods with a tough negotiating adversary.[9] We either have to give up these forms of communication or learn how to replace them through electronic media.

How much choice will we have? Fighting back against cocooning does not require luddite destruction of telephone centers, but it might require steadfast insistence on face-to-face contact, even when that is unpopular with efficiency-obsessed managers in the office.

We must recognize that major behavioral changes do not come easily, not nearly as easily as the installation of new hardware and computer software (which isn't all that easy either). I'm sure every person who has used a computer has experienced the fact that new software does not always work as advertised. How much harder it is to "program" the staff to cope with it!

Like it or not, though, if I had to make a prediction, it would have to be toward ever-increasing use of electronic media to displace in-person contact, and this will make us far less resistant to receiving our entertainment and education through electronic media.

Effects on Society and Industry

This part is why prediction is difficult. How will changes in technology create changes in human behavior? They always do. Anticipating how is the trick.

Effects of Leisure

Given today's multi-paycheck families and greater pressure on the job for more hours of work, often unpaid, we may have come to the point where no more free time is available. On the other hand, we also seem to be facing a permanently high rate of unemployment compared with what we have been used to. The unemployed will have more leisure, although that might not be an appropriate word for forced idleness. Governments may want to give them more media to keep them content with their lot, more or less what happened in Vonnegut's 1952 novel *Player Piano*, or for that matter in ancient Rome with the circus.

How much time we have for using media depends on what we use it for and what the competition is. If we do move toward a single medium, then it will be used a great deal for work and study as well as for leisure. In terms of entertainment, a move toward a single medium removes the fascinating question of which medium will win most of our time and attention. A Compaq television ad showed a young girl working on a computer and asking her mother what she did after school when she was young. The mother says she watched a lot of television. The girl, perplexed, asks, "Why?" The mother simply answers, with an unknowing shrug, "Good question." This ad is purely for one medium. It does not push any particular application, just the idea that a computer (Compaq, of course) is more interesting than television. Perhaps the more important question, for the industry as a whole, is how much time people will have for leisure activities. Could we come to see the day that the electronics industry, known for hard-driving employees, takes the lead in creating new leisure for the general workforce because they need it to sell their own products?

In the short run, time devoted to media can be only a little different from what it is now. It takes time to change habits and acquire new equipment. But as you watch drivers negotiating city traffic with half their attention on the cellular phone, or people walking down the street or attending baseball games doing the same, or people in airplanes and trains absorbed in their computers, it is pretty obvious that we are hooked on electronic media.

Effects on the Information Industry

The amount of telecommunications and number of computers in the home represent an enchanting dream for the companies that provide them. Is it possible that the rate of change will catch up even with them? Will such major investments as installing fiber-optic cable to the home or re-engineering the entire television system ever pay off before the next big change hits? The *New York Times* article referred to in Chapter 13 suggests these investments may not pay off. Profits on the Internet so far seem to be mainly accruing to the movers and searchers of information. Next in line are the searchers and retrievers. Will they take all of the revenue away from the providers of content?

There are a few companies making money selling information or entertainment on the Internet. There are a few more prospering by selling CD-ROMs for use in multimedia computers. John Seabrook in the *New Yorker* suggests that Microsoft may have peaked or is approaching its earnings peak.[10] If those small, communications-dependent computers come to be the dominant form of computing machine, how much of a market will there be for operating systems or expensive software, such as word processors, spreadsheets, or database managers? Maybe these same producers can make their money renting out time on large, remote computers using their software. On the other hand, Microsoft is still a new company. It has not been dedicated for a generation or more to a single way of doing things. It might be able to avoid the common corporate mistake of locking onto a single product line and technology and then, if the product market disappears, disappearing along with it.

Because time is limited, if we cut our hardware expenses in half, we cannot necessarily double the time we spend using computers. We can get rich without limit, but we cannot lengthen the day. The Lowry market model presented in Chapter 13 indicates that designs will have to change if content producers are to really improve their share of the entertainment and educational markets. Hence, the industry faces two conflicting demands: they are committed to a fast-obsolescence model and they need a market population with more leisure in which to consume their product.

Effects on Socioeconomic Balance

One of the favorite claims of technologists is that this marvelous new invention (whatever "this" is under discussion) is going to make life better for the working person, better educate the young, improve health through more information, etc. I wish I could predict something like this for the electronification of publishing. There will be benefits. Of that I am sure. But I do not see them leveling the socioeconomic playing field. It takes money to buy these things. The rich (middle class and above) have money and they do buy and use information technology. The poor do not on both counts. I cannot see information technology as a means of leveling society. If anything, the contrary.

I do believe those of us who do use information technology will have more data available to us than those who do not – anything from baseball scores to the best medical advice. We can get weather data before setting forth on a trip or gain knowledge about products we are thinking of buying. Will such a wealth of information improve our judgment? It should. It won't necessarily. Judgment requires understanding and more volume of data does not necessarily imply careful reading or understanding.

Politics seems a good example of this. We see less and less of our candidates for office on any one occasion, and more and more short sound bites. We get press releases and, in Canada and the U.K., flippant answers to questions posed in Parliament. Reasoned explanations are not much part of the game. Electronics has contributed to this state of affairs, as television appeal has come to dominate politics. Can this happen in other fields? Yes. We now have mass media advertising for doctors and lawyers. Will advertising come to dominate the choice of professional consultants as it does political candidates? It could.

Society, or its upper strata, will clearly become ever more media conscious and dependent, just as we became automobile dependent. Future society will, clearly, communicate extensively through the medium of electronic devices for interaction with other humans.

Chapter 17
Conclusions

I raised the question earlier of whether the continuing, huge industrial expenditures on infrastructure can possibly continue. This includes computers, computer software, video and cinema production, and telecommunications, as well as book and journal publication. So far, there has been great payoff in being a producer of computers and software, epitomized by the phenomenal rise of Microsoft. But as we switch ever more toward communication-based computing and entertainment, I wonder if the investments these changes require will pay off as well as they did for IBM and Xerox a generation ago, or Microsoft and Compaq in more modern times. Some observers believe that the late Cold War between the NATO and the Warsaw Pact countries was brought to an end because the Soviet Union could no longer afford the level of financial support required to sustain the arms race. Could that happen here, in telecommunications? I have to leave it as a question. I don't know the answer.

I cannot think of anything short of a major depression or a social catastrophe of equal proportions that is going to reverse or stop our relentless drive toward new media of communication. Therefore, I am quite convinced it will continue. This is going to mean that students of today will come to have a great many more experiences than those of us who were raised in a print, lecture, radio, TV, and cinema world. They will be able to accomplish more because any professional person will be able to consult with others or with reference works to an extent never before possible: doctors in small communities, lawyers facing specialized cases, real estate agents asked to find just the right accommodations for a client in a distant place. But all this will cost money.

Many, and perhaps eventually almost all, of these new experiences may be vicarious. We'll have some of the feeling of foreign travel or of dissecting a frog, but may never actually do either. As I have said before, I think this will be to the overall good of most people. They will learn more, and those with physical or emotional barriers to learning on one hand and those with great creativity and initiative on the other may learn far more. But the experiences will not be quite the same as the real thing, and we won't (I hope) want to license a surgeon who has worked only on virtual patients.

Will the people of this new world be better at developing deep insights or creativity? Or does it consist too much of pre-digested information, substituting the author's or artist's vivid presentation for the reader's or viewer's acts of creative imagination? Tom Harpur, a *Toronto Star* columnist, in an essay entitled "Information Highway Comes but Wisdom Lingers," put it that "our culture is so entranced by technology, consumerism and bottom-line thinking that the pursuit of wisdom has been practically obliterated from consciousness altogether."[1]

Yet, although we have no such thing as a census of great minds living at any given time, there is no reason to think we do not have our share of greats today. Of course, we would probably disagree on which ones of us they are. Only future people will be able to judge how we did in the pursuit of wisdom. Perhaps human adaptability is far greater than we give it credit for and, machines or no machines, we will continue to thrive, think, invent, and feel. If we can continue to do those things, that would not be a bad future for the human race. Some final comments follow on the form and content of future works, the mode of delivery, and the people who will use them, and then the ever-intriguing question of the survival of print.

Form

If publications are to become multimedia and electronically based, what will they be like? Trying to answer this question brings out a host of others. Will publications be just books or magazines recorded on discs, even if they are transmitted over the Internet for storage on our own discs? Will all books evolve into true multimedia works, complete with motion pictures and sound and, eventually, smell and touch? Will magazines and journals be distributed on the World Wide Web? If they are, will they retain the time-honored practice of being issued on a fixed date and with content fixed, once published? Or will a periodical become a publishing umbrella, under which individual pieces are published whenever they are ready? If a periodical is published on the web, on a publish-when-ready basis, for how long should the pieces be kept readily available to the reading public? Who would be responsible for keeping archival copies and making them available? Another way to look at this question is to ask if a journal is to become an ever-growing organism. If there is not always the concept of the "current issue," what, if anything, will replace it? How must we adapt our reading habits to a moving target, one that is or could be different each day? If it is different each time we access it, how will that affect what we have to pay for it?

The *New Yorker* magazine published a fascinating account of an experienced print editor who went to work at Microsoft to create and edit a new electronically based magazine.[2] He and his colleagues encountered most of the questions listed above and many more. There are no answers yet to most of them. Even Microsoft, an organization abounding with both creativity and money, cannot be sure how things will work out. We – the whole industry and its customers – are going to have to try things out and see what works, what is liked, what people will pay for, and what they will not tolerate.

Other than to say that documents of all sorts, whether term papers or books, seem to be moving toward use of more media, I find it hard to predict the exact forms they will take in the far future, dependent as that is on how people react to early experiments and trials.

Memory for machines is an important factor. When on-line database services were first initiated for the public, full text of documents was out of the question. That would have required more computer memory than was available. Throughout the history of computing, the solution to the need for more memory has always been to wait a year or two. For anyone who has used a personal computer since about 1990, just think of how memories have changed since then. The new DVDs will add to that, but they're not the end, just another step along the way. So, nothing that is prohibitive today because of insufficient memory will stay that way for long. On the other hand, always with more memory come more ideas on new information to store, leading to a need for more memory, leading to .... In brief, do not think of memory limitations as a limit on future growth.

Content

Form affects content. Pictures and sound can be used to some extent in place of words. I think we will see more of this. As for the real, underlying content, the substance, I am not sure there is any great call for change. We sometimes need assemblies of facts, sometimes descriptions of things, places, people, feelings, actions. These may be expressed differently, but we still need them.

Technology should enable us to avoid repetition, to allow, say, textbook writers to write only the parts they are really expert in and to rely upon others for parts of books they are expert in. The same goes for updating news items. Quick hypertext linkages to older material should reduce the need for endless repetition of the same facts. That is going to change the physical appearance of some publications. Less emphasis on a package, a bounded set of words and illustrations, more stress on connections to other information.

Repetition is not all bad. It is a valid pedagogical technique. What is desirable is to avoid unwanted repetition. There are other literary structures and techniques that will have to find their counterparts in multimedia if this new technology is to succeed.

Delivery

Here is where we'll see the most readily apparent change and it may reflect back on content and form. When delivery does not require transport of heavy objects, it is faster and cheaper. The speed can be dazzling. If you can get a copy of some obscure work from a library in a few seconds, the urge to own a paper copy of your own is lessened. That is an important economic consideration for publishers.

I just cannot see any serious obstacles to a situation where what we now think of as documents (books, pictures, musical recordings) can be transmitted quickly and without significant loss of resolution or fidelity, perhaps eliminating the need for or even convenience of individual ownership. Not tomorrow, but surely within a generation.

People

While I do not know if it has been definitely proven, it seems to me and to many others that we present-day people read differently from a computer screen than we do from paper. We scan rather than study on the computer. We are impatient with anything long. I, who will practically never write anything longer than a paragraph without a word processor, also will practically never read anything longer than a page without printing it first on paper. I suspect all this is because the electronic media are not really comfortable, so for the umpteenth time, I reiterate that the key to the transition is a good electronic reader. It is possible, also, that our impatience is not due solely to equipment deficiencies, but at least in part to our conditioning to believe that information coming from an electronic device (radio, TV, computer) *is* superficial and is not deserving of intense concentration or long study as is print. This applies to text, not numbers, which we might pore over for hours. The conclusion, then, is that either the form of text-based communication will change or we readers must change our habits.

Paying for Information

In North America we are used to libraries being free – free of most charges and of advertising. Our public schools are usually free, although free, public universities in our part of the world may have come to an end. Most government information used to be free, but our governments have caught on to the concept of cost recovery, meaning recovering the cost of publishing works from us, the taxpayers who paid for creating the works in the first place. Lately, even universities that have largely been funded by government have latched on to the idea of privatizing (read charging the full cost of tuition to students or their employers). Libraries are having trouble keeping up with rising costs of books, journals, and electronic equipment.

The Internet has made many people conscious of information availability and many users have come to believe it all is or should be free. We are used to paying money for newspapers and magazines. In many cases, broadcasters and publishers get more revenue from advertising than from subscriptions. Advertising as a means to have the consumer pay is interesting. If the ads did not lead to sales, the advertisers would not continue to pay for them. If we buy more than we need due to advertising, then we are paying in that way. If the ads are treated simply as public notices of goods or services available and we buy only what we need, the notices are closer to being free to us, although their cost is still bundled into the product cost. But it would have cost us something to find the information we needed through other means. Anyway, broadcast television and radio are "free," of all but advertising, after we have bought the receiving equipment. Will resistance to information for pay limit the spread of electronic publications? I think it may slow the spread but not stop it. Gradually, we will all grow accustomed to paying and then the resistance will disappear. It is the transition that hurts. Possibly, large-scale delivery of information for pay may lead to new government regulatory procedures as well.

It is to be hoped that people will continue to create new ideas rather than merely search files for evidence of existing ones. Again, only future historians will be able to look back and see how the generations following us did. It has been a theme in science fiction to picture a society in which all knowledge is known. There is no need to develop more, but there might still be a need to search for old knowledge.

The Disappearance of Print?

Will print survive electronics? I think we need to differentiate between the possible disappearance of print in the sense of ink on paper and its disappearance even in the form of electronically displayed shapes.

Books, letters, newspapers, and notes have enormous appeal to us today. To many of us nothing is as convient as carrying, reading, and possibly annotating a printed book. A pocket calendar is at least as versatile as an electronic one. But it seems almost foolish to insist that *never* can a line of computers be developed that can do what we want just as conveniently. I repeat: no machine currently on the market seriously threatens print. But, could it happen in ten, twenty, or thirty more years? Certainly. By that time another generation of computer users will have come along without as much resistance to the change as we book and paper people offer.

To me, the more interesting question is, "Could even the electronic form of print disappear?" That is, could even our computers stop communicating via keyboard and displays of alphabetic characters and deal entirely in pictures, the spoken word, and recognized gestures? Before trying to answer this question, let us consider why anyone might even want this to happen.

I cannot imagine anyone suddenly deciding to end print. What I can imagine is a *very* gradual increase in people's use of voice input and output and VR-like computer recognition of gesture. And I can imagine teachers and school boards *eventually* realizing that the effort to teach use of the written word is no longer worth it. There was a time, in Western society, when an educated person studied Latin and Greek. It was simply the thing to do. There was a time, far more recently, when an educated person studied geometry in secondary school with all the formality of postulates, theorems, and proofs. There was a time when at least a science-oriented student would have been expected to be skilled at mental computation. No more. Sometimes taught, rarely required. Many of us today regret these changes, but the world seems to have survived them and new skills, conflict resolution for example, are being taught instead.

One benefit of the disappearance of print would accrue to people born into a not-very-literate society. These people learn to speak, to listen, and to understand language. But they may not learn to read or write. Their lack is of schooling, not of brains. They are virtual outcasts in the modern world. But do we who are literate want to give up our precious print to help the less fortunate? The reality is almost certainly that we do not. But that is today. When school curricula are so crowded with new subjects, and teachers recognize the superfluity of print, will the people of that day feel so strongly about it? They will when it is first brought up, but the resistance may disappear.

Would people miss print if it disappeared from common use? I think not. I think the change would be so gradual that it would not matter to most. Would all books disappear? No. They would move into museums and the occasional artisan would produce one to show how something once important was done. This is what happened to clay tablets, scrolls, and Gutenberg books.

Would not the speed of communication suffer? That is, can we not read much faster than we talk? People read, or can read at several hundred words per minute (wpm). Speed readers can go much higher. Many cannot get much beyond 100. A lecture might be delivered at about 100 wpm but an announcer's description of a play in football goes much faster. We can hear and comprehend around 200 wpm but it takes effort over a long period of time. There has been machinery developed that can speed up recorded speech by a method that speeds up the delivery of a word and also deletes much of the silence between words. It takes some learning to work with this form of speech, just as it takes learning to speed read. The point is that we would not necessarily have to give up speed in communication between humans and computers or humans and whatever replaces the written word.

This prediction is one of those I consider useless in one sense because no one reading it now will see the day it comes about. And if it does not come about, my future defenders can simply say, "Wait 'til the next millennium." But the value of the prediction is to encourage people to do some serious thinking about media, what we want from them, what aspects are really important. Is the medium more important than the message?

Summary of Conclusions

Changes will occur. Some will be seen as good, some bad, and some neutral. Unfortunately, which outcome gets which rating is not something many of us are likely to agree on. Further, the good and bad are not mutually exclusive. Some developments could bring both outcomes. Here are my list and ratings:

Good Things

A great variety of information products and services will be possible.

The lowering of communications barriers across geographical boundaries will continue to occur and we might see a crossing of emotional ones as well, as people "meet" the sort of people they might never encounter face-to-face.

There will be an opportunity for more individualized learning, keyed to abilities, interests, and motivations of students, resulting in a significant leap forward in what future people will earn

There will be removal or lowering of arbitrary restrictions that limit authors or creators from getting their work to the public. These are principally fiscal, but may also be political.

Technology will continue to change, enabling us to accomplish new things and to support an industry that depends on rapid technological obsolescence.

Neutral Things

All current classes of literature (prose, poetry, photography, cinema, etc.) will make ever more use of multiple media.

The need for print will gradually disappear over many, many years. (Neutral? Yes, for reasons stated previously, mainly that the people of the time will not see it as a tragedy.)

Bad Things

More technology costs more money, and even if we find the money, it doesn't always work. Therefore, the blessings of a high-tech world will not accrue equally to all, or even nearly so.

Just as the Cold War ended, some think, because one side could no longer afford the payments, we may choke on technical obsolescence.

Overconcentration of the major corporate producers of information products and services can lead to more uniformity and a lessening of creativity. "Corporate" is not limited to for-profit institutions.

The ease and fun of browsing and retrieving what others have produced could reduce not only the creativity and level of comprehension of students (future adults), but ultimately reduce societal recognition of their importance.

There will be removal or lowering of arbitrary restrictions that limit authors or creators from getting their work to the public. These are principally fiscal, but may also be political. (Yes, this also appeared under *Good Things*. Clearly, its classification depends on what is published and who is reading.)

The very final conclusion is that electronic publishing is not something to fear. It has at least as much promise as threat. It does, however, call for some intelligent thought before jumping into the next glib announcement from the marketers. But then, what doesn't?

Appendix

This appendix contains statistical charts, all of which are time series showing how various factors relating to electronic publishing have changed over the years. Hence, in most graphs, the horizontal axis represents *year*. Read carefully what the variable reported is. These can be confusing. For example, it is a common supposition that newspaper circulation or readership is declining. But what appears to have happened is that many daily papers have been discontinued or merged into other papers. New ones, often specialized by subject or neighborhood, have cropped up. As a whole, the industry seems to be surviving nicely.

The charts give a flavor of the growth of the communications industry, in general, and the print sector in more detail. Some were directly referred to in the text, others are provided here for the reader to browse or study. All the data are for the United States. While there has been some criticism of the 10-percent rule for Canadian data, namely that any Canadian economic data relating to population or markets are 10 percent of the U.S. values, in the data shown here, this supposition tends to hold within a few percent variation.

Some of the data, such as gross national product, are authoritative and presumably reliable. Some, such as number of books published world-wide, are quite difficult to come by. In modern times that figure is col-lected by UNESCO, which is, in turn, dependent on individual countries to provide them with data. Reporting is haphazard, hence the totals are seemingly, at best, rough approximations. But our aim is mainly to show trends and these come out quite clearly in most cases.

Another problem is that sometimes we have drawn upon more than one source, typically one source for older data, a different one for more modern data. Sometimes the two do not combine seamlessly. One exam-ple is in the number of telegraph messages handled per year (Figure A12). The older data is for the Western Union Telegraph Company, once the dominant firm in the industry. Later data is industry-wide. There is a gap between the two series just prior to which Western Union rates begin to decline. This is probably indicative of competition entering the market, eroding Western Union's dominance, before the telephone began to dis-place the telegraph in general.

The reader is advised not to take these data definitively (I am uncer-tain about the reliability of some of the source material), but definitely to consider the trends as reliable. I found many of these fun to browse through and I encourage readers to do the same. Most graphs contain more than one data series. The listing below shows the series and the charts in which they are contained.

Series	Figures

1. Demographic Data

Population	A1
Number of Households	A1
Population Projections by Age	A2, A3

2. Economic, Educational, and Cultural Data

Gross National Product (GNP)	A4
Family Income	A5
Personal Expenditures on Recreation	A5
Degrees Conferred	A6
Copyrights Registered	A6
Patents Issued	A6

3. Print and Disk Publication

4. Nonprint Media

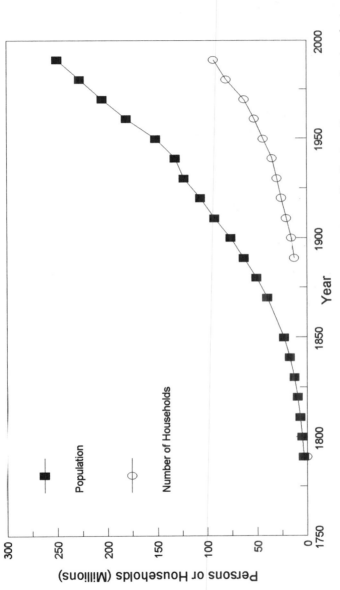

Figure A1. U.S. population and number of households. Many figures on usage or distribution of communication devices or services are given in terms of these variables. Sources: Kurian, Datapedia, series A6, A335, Statistical Abstract 1996, Tables 16, 66.

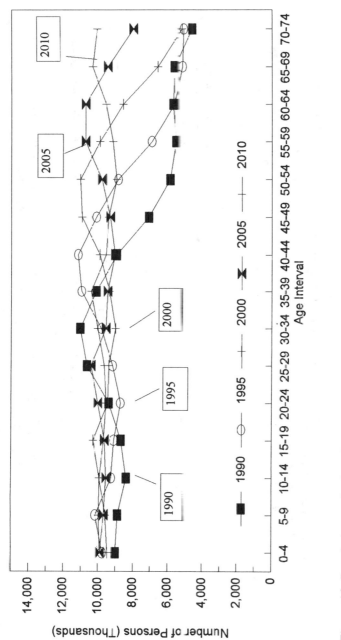

Figure A2. Population projections by age. *The expected number of persons in various age groups for years 1990 through 2010. Source: Bos et al., World Population, p. 494.*

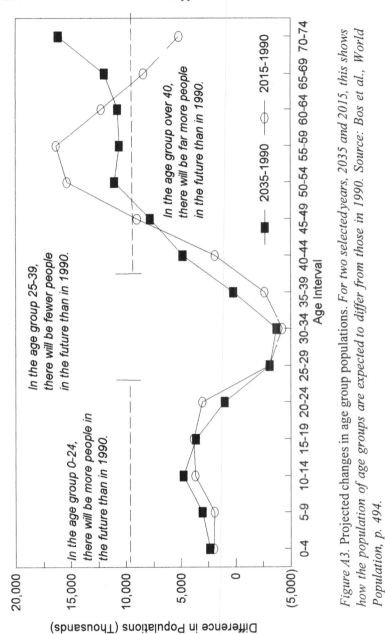

Figure A3. Projected changes in age group populations. *For two selected years, 2035 and 2015, this shows how the population of age groups are expected to differ from those in 1990. Source: Bos et al., World Population, p. 494.*

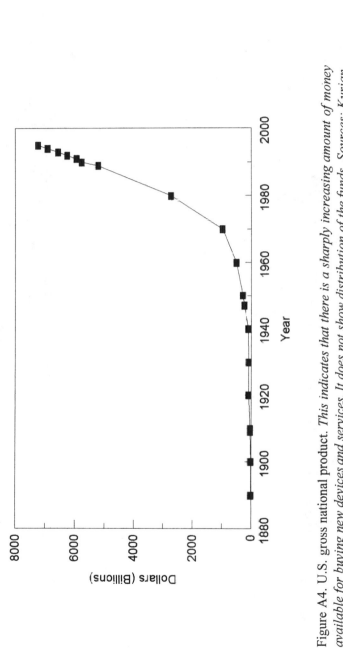

Figure A4. U.S. gross national product. *This indicates that there is a sharply increasing amount of money available for buying new devices and services. It does not show distribution of the funds. Sources: Kurian, Datapedia, series F1, G179; Statistical Abstract 1996, Tables 16, 709.*

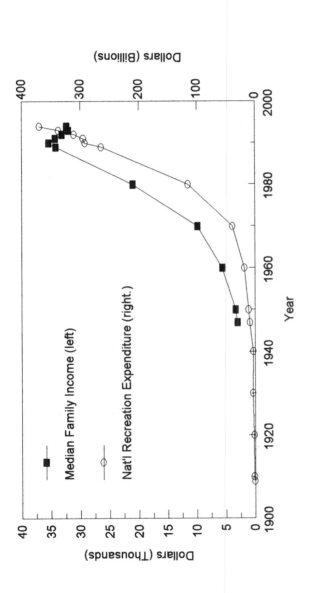

Figure A5. Family median income and total expenditures on recreation. *Average family income had been rising much like GNP, except for the most recent few years. Expenditures on recreation continue apace. Note that the recreation figures are totals for the U.S.A., not averages per family. Sources: Kurian, Datapedia, series H878; Statistical Abstract 1996, Table 401.*

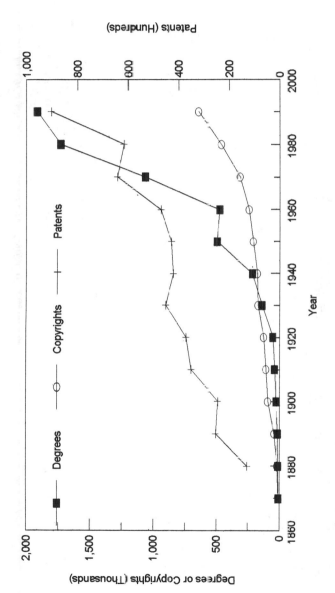
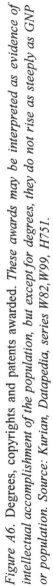

Figure A6. Degrees, copyrights and patents awarded. *These awards may be interpreted as evidence of intellectual accomplishment of the population, but except for degrees, they do not rise as steeply as GNP or population. Source: Kurian, Datapedia, series W82,W99, H751.*

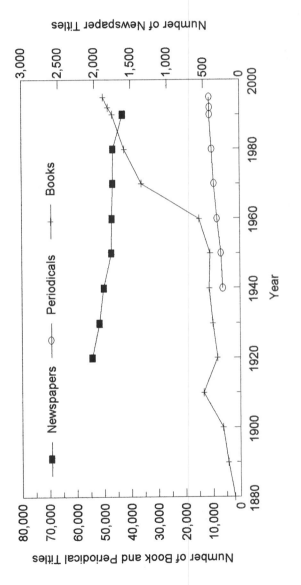

Figure A7. Number of newspapers, other periodicals, and books published, by title. *Newspaper titles are declining somewhat; periodicals are on a slow increase, but books seem to be growing rapidly in number. Sources: Kurian, Datapedia, series R224, R237, R192; Statistical Abstract 1996, Tables 896, 900.*

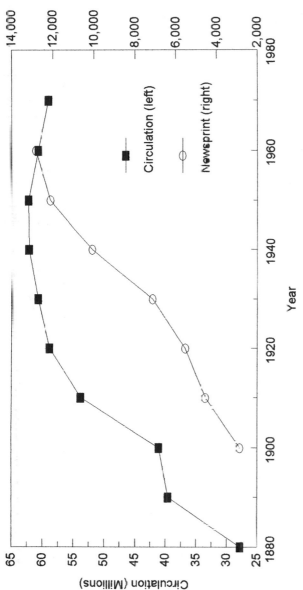

Figure A8. Newspaper circulation and newsprint consumption. *This is another way to judge how newspapers are doing. Total circulation is declining slightly, but the amount of paper consumed has been going up, although we do not have figures for the latest years. Numbers of titles are more indicative of reader interest than of reader purchases. Source: Kurian, Datapedia, series L194; Statistical Abstract 1996, Table 899.*

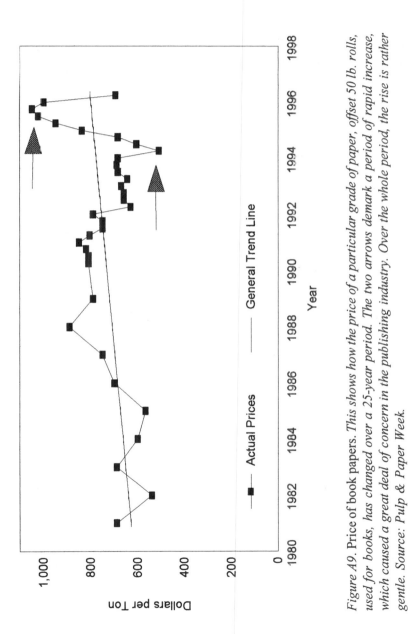

Figure A9. Price of book papers. *This shows how the price of a particular grade of paper, offset 50 lb. rolls, used for books, has changed over a 25-year period. The two arrows demark a period of rapid increase, which caused a great deal of concern in the publishing industry. Over the whole period, the rise is rather gentle. Source: Pulp & Paper Week.*

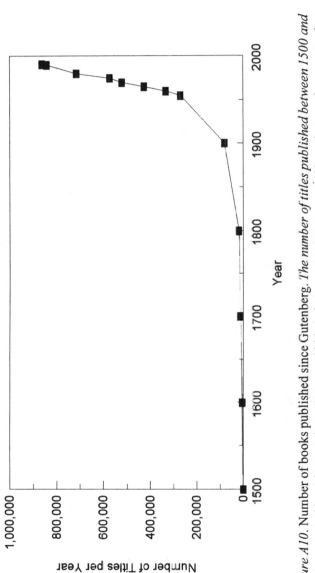
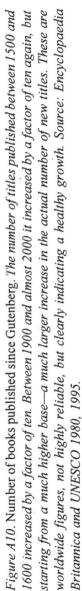

Figure A10. Number of books published since Gutenberg. *The number of titles published between 1500 and 1600 increased by a factor of ten. Between 1900 and almost 2000 it increased by a factor of ten again, but starting from a much higher base—a much larger increase in the actual number of new titles. These are worldwide figures, not highly reliable, but clearly indicating a healthy growth. Source: Encyclopaedia Britannica and UNESCO 1980, 1995.*

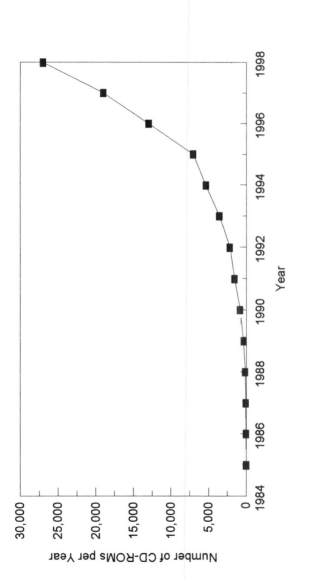

Figure A11. Number of compact disk publications. *Another nicely rising growth curve. Because of the lack of clear definition of what constitutes a publication, it is difficult to be sure how many of these represent what might be termed "books." Source: Multimedia and CD-ROM.*

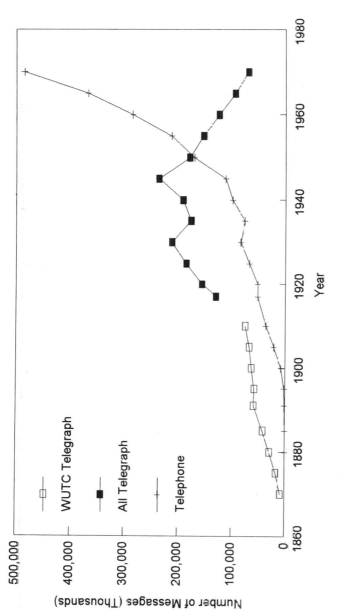

Figure A12. Traffic volumes for telegraph and telephone. *This graph shows when the telephone took over from the telegraph, the latter an industry that has now virtually disappeared, a rare case of one medium totally displacing another. Sources: Historical Statistics, series R46, R56; Statistical Abstract table 882.*

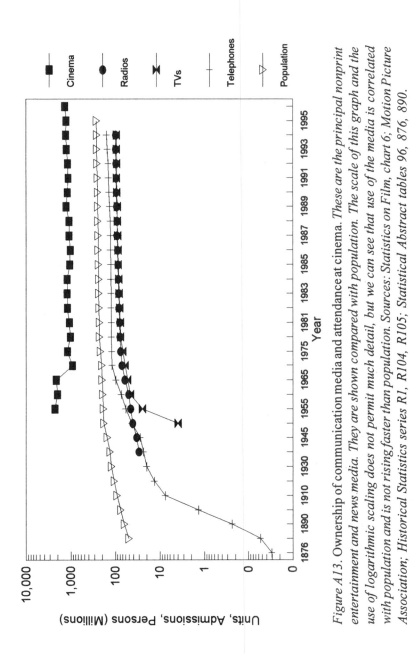

Figure A13. Ownership of communication media and attendance at cinema. *These are the principal nonprint entertainment and news media. They are shown compared with population. The scale of this graph and the use of logarithmic scaling does not permit much detail, but we can see that use of the media is correlated with population and is not rising faster than population. Sources: Statistics on Film, chart 6; Motion Picture Association; Historical Statistics series R1, R104, R105; Statistical Abstract tables 96, 876, 890.*

Figure A14. Use of telephone and television, via wire or cable and via cellular or satellite. *While wire-based telephone service continues to grow, the rate of increase is slowing down. The rate of increase of cellular-PCS users is much steeper. Similarly, television via satellite is beginning to make inroads on the cable. Both the figures for the year 2000 are optimistic industry projections. Will we see the same pattern here as we did in A12 for telephone and telegraph? Sources: Historical Statistics series R1; Statistical Abstract tables 882, 884, 876; Hart; U.S. Economic Review, p. 15.*

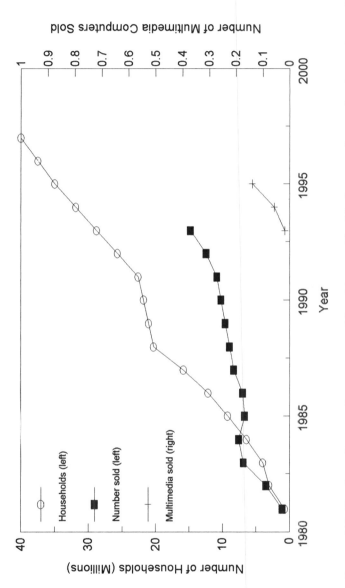

Figure A15. Number of personal computers sold and in use. *The black rectangles mark sales per year of personal computers. The hollow ovals show number of households having a computer. The crosses show the number of multimedia computers in use. Sources: Statistical Abstract 1996 table 1234; DISCIS Knowledge Research, Inc.; Weidt.*

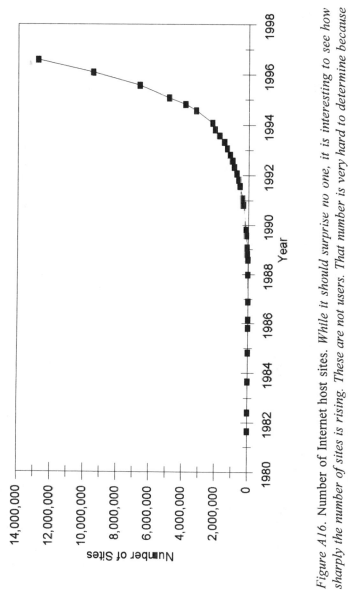

Figure A16. Number of Internet host sites. While it should surprise no one, it is interesting to see how sharply the number of sites is rising. These are not users. That number is very hard to determine because so many people make use of school or office machines. Source: SRI.

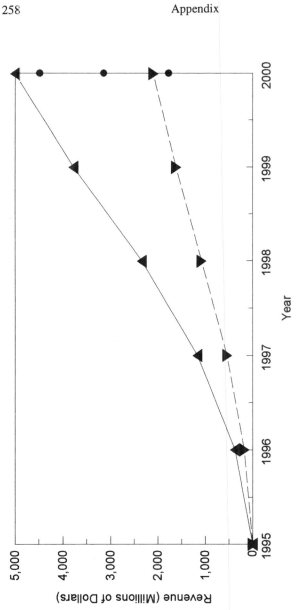

Figure A17. Revenue from advertising on the Internet. *This shows a composite of several estimates and projections. The upper line is an optimistic estimate, the lower from another, more pessimistic source. Both show an increase, but not as steep as that for sites. The black dots on the line for 2000 are yet other estimates for that year only. Source: Toronto Globe & Mail, internal statistics from a variety of their sources.*

Notes

Citations within these notes are to the list of cited works in the bibliography that follows.

Chapter 1

1. Meadow and Yuan, 1997. Quoted in Duff, Craig, and McNeil, *Note of the Origins*, 117.
2. Quoted in Zable, *Jewels*, 209.
3. McLuhan, *Understanding*, 9.
4. See Ruby, "Terminal Behavior," for a quick summary of some research and Kiesler, Siegel, and McGuire, "Social Psychological" for a more formal report of the research.
5. An interesting and brief history of the development of science publishing and newspapers is found in Marten, "Developing of European."
6. Some of this is only vaguely remembered by me, some reported in Wood and Wylie, *Educational.*
7. Bolter, "Literature."
8. Weizenbaum, *Computer Power,* p. 207.

Chapter 2

1. Shannon and Weaver, *Mathematical Theory*, 4-6. The title and much of the content are daunting for the nonmathematical reader, but the part written by Weaver (95-117) is fairly readable and fascinating.

2. Meadow, *Text Information,* 20-31.

3. Shannon and Weaver, *Mathematical Theory*, 99-106.

4. Boorstin, *Gresham's*, p 3.

5. Hayakawa, *Language*, 60.

6. Cherry, *Human Behavior*, 10.

7. Negroponte, *Being Digital* and Eisenhart, *Publishing.*

Chapter 3

1. Chomsky, *Reflections* and Pinker, *Language Instinct.*

2. Quoted in Jean, *Writing*, 83.

3. John Shaw Billings, an American Army medical officer with an interest in vital statistics, was put in charge of the 1890 U.S. Census. He is said to have suggested rather casually to Herman Hollerith, a member of his staff, that data might be recorded on punched cards. Billings did not follow up. Hollerith developed punched cards and the machines to use them and went on to found a company that, upon merger with another, became IBM. See Sobel, *IBM.*

4. Duggan, *God and My Right*, 150-151.

5. Augustinus, *Confessions*, 84.

6. Illich, *Vineyard* and Crowley and Heyer, *Communication.*

7. Levarie, *Art & History*, 78-84.

8. Kunhardt, *Pat the Bunny.*

9. Drucker, "Infoliteracy."

10. *Statistical Yearbook*, 179.

11. *The Canadian Global Almanac* and *Statistical Abstract of the United States.*

12. Mosco, *Transforming.*

13. Lewis, *Empire*, 105-107.

14. Camp, "Misinformation."

15. O'Neill, "Role of ARPA," and Quatermain and Hoskins, "Notable Computer Networks."

Chapter 4

1. Burckhardt, *Civilization*, 204.

2. Chambers, "Difference." (1993).

3. Barthes, "Work to Text."

4. Bolter, "Literature in the Electronic."

5. Nelson, "Opening."

6. Illich, *Vineyard*, 15.

7. Illich, *Vineyard*, 104n.

8. Prior to the start of the first trial the presiding judge ordered that there be "no publication of the circumstances of the deaths of any persons referred to during the trial and they shall not be revealed directly or indirectly to a member of the foreign press," as reported in the Toronto *Globe and Mail*, 7 July 1993, A4. Only members of the Canadian news media were admitted to the court. This was done to protect the rights of the second accused to a fair trial. But those of us with access to electronic news sources from the U.S.A. lacked no information. Later, when the second trial was held, at the same time that the Simpson trial was held in Los Angeles with all its full public disclosure of the details, many of us appreciated the relative calm of this trial compared with that in L.A. The lack of the circuslike atmosphere seemed to add to the stature of the court.

9. Boorstin, *Gresham's*, 3.

Chapter 5

1. Negroponte, *Being Digital* and Eisenhart, *Publishing*.

2. Pentland, "Smart Rooms." For some technical summaries of work being done see: Flickner *et al.,* "Query by Image," Moghaddam and Pentland, "Face Recognition," and Cardenas *et al.,* "Knowledge Based."

3. But with clean, sharp input text, modern scanners and OCR software give excellent results.

Chapter 6

1. Pinker, *Language Instinct.*

2. Oates, *Black Water.*

3. The memex paper is Bush, "As We May Think." A history of his work is found in Nyce and Kahn, *From Memex.*

4. Englebart, "Conceptual Framework" and Nelson. "Opening Hypertext."

5. Englebart and English, "A Research Center."

6. Alpert and Bitzer, "Advances." This and the citations in note 7 are old discussions of old technology, but still of interest.

7. Hartley, "Programmed Instruction" and Mackie "Programmed Learning."

8. For writings about hypertext, see: Englebart, "A Conceptual Framework" Nelson, "Opening Hypertext," Blumberg, "Ex Libris," and Tuman, *Literacy Online.* For examples of hypertext books see the following web sites:

   http:/www.mshumate.duke.edu/ and

   http:/www.eastgate.com.

9. An excellent brief review of some of these was written by Burkhard Bilger, "World on a Platter."

10. Liddy, *et al.* 1995.

11. Miller, "WordNet: an online" and "WordNet: a lexical."

12. Nelson, "Opening Hypertext."

13. Blumberg, "Ex Libris."

14. See Bolter, "Literature in Electronic," referring to Joyce's (*Afternoon*), hypertext short novel, published on a floppy disc.

15. Bolter, "Literature in Electronic," 29, 30.

16. Logan, *Fifth*, 57.

17. See the newspaper article by Hilkevitch, "CD-ROM" and a review of two frog dissection kits by Sweitzer, "Slice."

18. Many people have heard one version or another of this. I have heard it attributed to Maria Montessori, a Chinese proverb, and a Native American proverb. This last source was suggested by Julie Levang at the University of Michigan. It doesn't matter. It's good thinking.

19. Marchionini, *Information Seeking*, 114-117, and Bates, "Design of Browsing."

Chapter 7

1. de Hartog, *Captain.*

2. Hodge, *Interactive Television.*

3. Licklider, "Man-Computer."

4. Rheingold, *Virtual Reality,* and Biocca and Meyer, "Virtual Reality."

5. In June 1995 Microsoft introduced a multimedia version of *How the Leopard Got His Spots* narrated by the actor Danny Glover. At the same time, they introduced a new CD-ROM story disc, which they plan to distribute free to 7000 children's librarians and which, according to their press release, "integrate[s] multimedia technology with children's storytime, one of the most important services our libraries provide."

Chapter 8

1. Quoted in Soloway and Pryor, "Next Generation."

2. Aston and Schwarz, *Multimedia,* Earnshaw *et al.,* *Digital Media,* and Neilsen, *Hypertext.*

3. See, for example, "Spielberg's Lament" and Blum, "Emotion Pictures."

4. A good, nontechnical introduction to the issues of people relating to machines is Donald Norman, *The Psychology of Everyday Things.* A more technical treatment is Shneiderman, *Designing.*

5. Paul, "Common Sense?"

6. There is quite a bit of poetry on the web. A few examples, with their web addresses, are:

Edna St. Vincent Millay, a collection of her poetry assembled at the University of Maryland: http://www.inform.umd.edu:8080/EdRes/Topic/WomensStudies/ReadingRoom/Poetry

Hear Poetry, Best Quality Audio Web Poems, a raucous collection from Brown University that features the works being presented to the audience in audio form, not as images of printed words: http://www.cs.brown.edu/fun/bawp

Intimations, poetry by David Miller illustrated with photographs: http://www.awa.com/w2/intimations/i-1.2.html

Interactive Poetry Archive, a collection of published poetry, featuring photographs and the user's option of reading or hearing the works: http://sunsite.unc.edu/ipa

7. Sengstack, "DVD."

Chapter 9

1. Gurrie and O'Connor, *Voice-Data*; Blyth and Blyth, *Telecommunication*; and for telecommunications law Kennedy, *Introduction*.

2. Stoffels, "Carter," and "Thomas F. Carter v".

3. Clark, *Networks and Telecommunications*, pp. 297-318, 311-312.

4. Clarke, *How the World Was One*.

5. There is a huge printed literature on the Internet and its web. Here are a few. Most are issued in new editions relatively often. Some examples are: Carroll and Broadhead, *Canadian Internet*, December and Randall, *World Wide Web*, *Internet Unleashed*, Krol, *Whole Internet*, and Nielsen, *Hypertext*.

6. Berners-Lee, "WWW: Past, Present, and Future."

7. Shapiro, *Collaborative Computing*.

8. Gore, "Networking." In this article U.S. Vice President Al Gore, then a senator, referred to his proposal "eleven years ago" for a network of "data highways." In the ensuing years, the prefix *super* was added. The Vice President's father, when he was in Congress, was a sponsor of the legislation to create the U.S. Interstate Highway system, so perhaps that is where the idea of superhighways came from.

9. Donald Case, "Rhetoric," discussed some of the confusion over names of networks.

Chapter 10

1. Bagdikian, *Information Machines*, 3.

2. The War of 1812 ended on December 24, 1814, with the signing of the Treaty of Ghent. The Battle of New Orleans took place on January 8, 1815. There was no way to move information from Ghent, Belgium, to New Orleans fast enough in those days to prevent the battle. Tragic as it was for those who did

not survive, it eventually led the victor, Andrew Jackson, to the U.S. Presidency.

3. The battle effectively ended on June 18, 1815. A Rothschild agent simply bought a Dutch newspaper with an account of it and took the first boat to England, arriving on June 20. He beat the official news by one day. A legend grew up that Rothschild's people used carrier pigeons, but they simply made the best use of public information and public transportation, according to Morton, *Rothschilds*.

4. Young, "Classes."

5. Brehl, "Big Fight," and "Bell rattles the net."

6. "Prestel."

7. Ogden, *Last Book*.

8. Barlow, "World Trends."

9. There is an excellent summary of the "discovery" and announcement of cold fusion in a book review by Crooks, "Bad Science" and the problems that hasty, unrefereed publication created. The book reviewed is Taubes, *Bad Science*.

10. See Harrison and Stephen, "Electronic Journal," a lengthy and somewhat technical article; Okerson, "Scholarly Publishing," a collection of articles; Frankel, "Daily Digital," a brief article on electronic newspapers; Savetz, "Magazines." a brief article on electronic magazines; and an excellent statement by Peter Denning and Bernard Rous, "ACM Electronic," about scientific journals in general and the position of the Association for Computing Machinery on them.

11. Quoted in Krumenaker, "Virtual Libraries."

12. Cronin and Overfelt, "E-Journals," and Leslie, "Goodbye Gutenberg."

13. Maran, "Science," recounts some tales of premature, early, and late announcers of scientific discoveries.

14. Snoddy, "McGraw-Hill," and "Publisher Who Had a Global Electronic Dream."

15. See Zijlstra, "University Licensing." The Academic Press electronic journal system is called IDEAL. Information can be found on the Internet at http://www.idealibrary.com.

16. Taylor, "Textbooks" and Milliot, "New Media."

17. LaFollete, "Stealing," and Hammerschmidt and Gross, "Biomedical Fraud."

18. Lyman, "What is a digital," and Okerson, "Who Owns"

19. Feldman, "DIALOG."

Chapter 11

1. I bring this subject up as a negative example of measurement of complex social phenomena. Among the most famous race and intelligence works are

those of Jensen, "Race, Intelligence," and "Race and Mental." More recently Charles Murray's and Richard J. Hermstein's *The Bell Curve* created quite a furor and an equally large one was created in Canada by Prof. Phillippe J. Rushton who compared blacks, whites, and Asians on a number of bases including cranial capacity and genital size. See Jones, "Canadian's Genetic Theories," for a newspaper account and Rushton *et al.*, "Race Differences" and Genetic and Environmental" for the details. Some works that tend to refute the relationship between race and intelligence or the difficulty of measuring intelligence are: Fraser, *Bell Curve Wars*, Loring *et al., Race and Intelligence,* and Senna, *Fallacy.*

2. Samuel Johnson, quoted in *Oxford Dictionary of Quotations.*

3. Blumberg, "Ex Libris."

4. Some research at Drexel University by Mancall and Drott, "Tomorrow's Scholars," showed that more than half of high school students assigned to do research papers used their home library as a source. This included students from inner-city schools whose location typically would mean not much money available for a serious library. But most found something useful. And this was before home computers became popular. With computers now in something like 36 percent (and growing) of North American homes, the number of potential sources is exploding.

5. Blackwell, "Imagining Things."

6. Harel and Papert, *Constructionism*, and Harel, *Constructionist.*

7. This is an issue beginning to worry more and more people. Some discussions from the legal point of view are found in Tarter, "Information Liability," and Watson, "Liability."

8. Drucker, "Infoliteracy."

9. I keep meeting people who believe this, and I see it written in print, but I have yet to find a definitive study that proves it is true So we must consider it a theory, unproved.

10. Blumberg, *op cit.*

11. Elizabeth R. Kolasky, personal communication.

Chapter 12

1. Rogers, *Diffusion.*

2. Data on paper prices is from *Pulp & Paper Week*, published by Miller Freeman, Inc., and provided to me by Mirjana Risek of that company.

3. Negroponte, *Being Digital.*

4. Brehl, "Big Fight."

5. "Who pays for it?" in Krol, *Whole Internet*, 17.

6. The real costs of conversion are very difficult to determine in advance. Lesk, "Going Digital," offers an analysis of the issue and some references for more research.

Chapter 13

1. The data for this graph and the others to follow in this chapter are generally composites, mostly taken from variations of the United Nations' *Statistical Yearbook* and of the *Statistical Abstract of the United States*. It became clear as I went through the process of gathering the data that estimates are often inconsistent. Please regard the graphs as general indicators of trends, not as firm measures of what happened in any given year.

2. Jenkins, "Death of the Written."

3. Production Screen Life, *End of Television*.

4. Archer, "CD-ROMs."

5. Maughan, "Whither multimedia?"

6. Gates *et al.*, *Road Ahead*, 119.

7. John Lowry, personal communication to the author, 1996.

8. There have been many mentions in the press of the possibility of such computer-communications systems. An example is: "IBM announces `net centric' computer," (so listed in the Bibliography).

9. Levy, "How the propeller."

10. See "Report of the Commissioner," Sword, "Photocopying," and Yanow, Copying." In brief, Williams and Wilkins, a private publisher of scientific journals, sued the National Library of Medicine for copyright infringement because of the Library's policy of making copies of journal articles for members of the medical profession. After a preliminary hearing, court hearing, and Supreme Court review, the Library (officially the United States) won, by a squeak.

Chapter 14

1. The Telecommunications Act 1996, P.L. 104 104, 104th Congress, also known as the Telecommunications Decency Act.

2. Stets, "Law Experts," describes the first rejection by a Federal court and Greenhouse, "Supreme Court," discusses the Supreme Court action.

3. Meyer, "A Bad Dream," and Campbell, "Censoring."

4. Wilson, "Northern Exposure."

Chapter 15

1. "Arthur Conan Doyle's Sherlock Holmes available on CD-ROM."

2. Gates, *Road Ahead*, 91.

3. As reported by the Associated Press, "Computer Wins."

4. Manguel, *History of Reading*, 318.

5. Nelson, "Opening Hypertext."

6. Khayyam, *Rubaiyat*, 58.

Chapter 16

1. Clarke, *How the World Was One*, 113.
2. Grosch, *Computer*, 131, Lenzner, "Reluctant."
3. McLuhan, *Gutenberg Galaxy*, Inn s, *The Bias of Communication*, Logan, *The Fifth Language.*
4. I do not know the exact source of this. I turned a television on one day in August 1997 and saw an interview with Howard already in progress. I do not even know what channel or network I was watching.
5. The continuing increase in collection size is not in conflict with the prediction of decreased variety. Just as with Hollywood movies we may see more instances of fewer types of work
6. I've been finding, at least in graduate school, that scheduling meetings among students is no small problem. More and more, our students have jobs or families, and may live a long way from the campus. E-mail is a great blessing for them.
7. Drucker, "Infoliteracy."
8. Terkle, *Second Self.* This excellent book describes the "computer hacker" phenomenon. Also, there have been letters in Ann Landers's newspaper advice column reporting that a husband or wife has left his or her spouse to take up with a person known only through electronic mail -- someone never met in real life.
9. The reference here is to the 1982 disarmament negotiations between the United States and the Soviet Union. At a tense, stalemated point, the respective delegation heads, Paul Nitze for the U.S. and Yuri Kvitsinsky for the U.S.S.R., went together, but privately, for a walk in the woods. They were able to break the deadlock by a more personal approach. Could this have happened in a machine-mediated proceeding?
10. Seabrook, "Gates of the Temple," and Hertzberg, "All-out Push."

Chapter 17

1. Harpur, "Information Highway."
2. Auletta, "Annals of Communication."

Bibliography

Alpert, D., and D.L. Bitzer. "Advances in Computer-based Education." *Science* 167 no. 3925 (1970): 1582-1590.

Archer, Bert. "CD-ROMs: The Affair Is Over." *Toronto Globe and Mail, Report on Business Magazine*, 31 May 1997: 68.

"Arthur Conan Doyle's Sherlock Holmes Available on CD-ROM." *PC Magazine* 11 no. 11 (1992): 481 et seq.

Aston, Robert, and Joyce Schwarz. *Multimedia: Gateway to the Next Millennium*. Boston: AP Professional, 1994.

Augustinus, Aurelius. *The Confessions of St. Augustine*. New York: P.F. Collier & Son, 1909.

Auletta, Ken. "Annals of Communication: The Reeducation of Michael Kinsley," *The New Yorker* (May 13, 1996): 58-73.

Bagdikian, Ben H. *The Information Machines*. New York: Harper & Row, 1971.

Barlow, D.H. "World Trends in Physics Information Systems." *Physics Bulletin* 23 (Sept. 1972): 519-20.

Barthes, Roland. "From Work to Text." In *Textual Strategies: Perspectives in Post-Structural Criticism,* edited by J. V. Harari. Ithaca: Cornell

University Press, 1979, 73-81.

Bates, Marcia. "The Design of Browsing and Berrypicking Techniques for the Online Search Interface." *Online Review* 13 no. 5 (1989): 407-424.

"Bell Rattles the Net." *Toronto Star,* 9 November 1995, D-1.

Berners-Lee. "WWW: Past, Present and Future." *Computer* 29 no. 10 (1996): 69-77.

Bilger, Burkhard. "World on a Platter: The Short Happy Life of Electronic Encyclopedias. *The Sciences* 37 no. 4 (July/August 1997): 42.

Biocca, Frank, and Kenneth Meyer. "Virtual Reality," in Aston and Schwarz, op. cit., 185-228.

Blackwell, Gerry. "Imagining Things." *Toronto Star, Fast Forward,* 8 June 1995, J6.

Blum, David. "Emotion Pictures: Directors Cry about Colorization." *New Republic,* 196 (1987): 13-15.

Blumberg, Roger B. "Ex Libris." *The Sciences* 35 no. 5 (1995): 16-19.

Blyth, Mary M., and W. John Blyth. *Telecommunications: Concepts, Development & Management.* New York: Macmillan, 1984.

Bolter, Jay David. "Literature in the Electronic Writing Space," In *Literacy Online.* Edited by Myron C. Tuman. Pittsburgh: University of Pittsburgh Press, 1992, 19-42.

Boorstin, Daniel J. *Gresham's Law, Knowledge or Information,* Remarks at the White House Conference on Libraries and Information Services. Washington: Library of Congress. Center for the Book, 1980.

Bos, Eduard, M.T. Vu, E. Massiah and R.A. Bulatao. *World Population Projections 1994-95 Edition* Baltimore: Johns Hopkins University Press, 1994.

Brehl, Robert. "Big Fight Looms for Highway Rider." *The Toronto Star,* 6 November 1995, C3.

Burckhardt, Jacob. *The Civilization of the Renaissance in Italy* vol. 1. New York: Harper & Row, 1958.

Bush, Vannevar. "As We May Think." *Atlantic Monthly* 76 no. 1 (1945): 101-108.

Camp, Dalton. "Misinformation Travels Fast on the Information Highway." *Toronto Star.* July 26, 1995 A17 Reprinted with permission — courtesy The Toronto Star Syndicate.

Campbell, K.K. "Censoring the Net a Pointless Pursuit." *Toronto Star,* 8 February 1996, G3.

The Canadian Global Almanac. (Toronto: Macmillan Canada, 1963)

Cardenas, A.F., I.T. Ieong, R. Barker, R.K. Taira, and C.M. Breant. "The Knowledge-based Object-oriented Picquery+ Language." *IEEE Transactions on Knowledge and Data Engineering* 5 no. 4 (August 1993): 644-658.

Carroll, Jim, and Rick Broadhead. *Canadian Internet Handbook.* Scarborough ON: Prentice-Hall Canada, 1996.

Case, Donald O. "Rhetoric of the 'Ren: the Myths of the Information Surperhighway," *Proceedings of the 58th Annual Meeting of the American Society for Information Science,* Vol 32. (Medford, NJ: Information Today, Inc., 1995) 137-143.

Cellular Telecommunications Industry Association, Web site http://www.wow-com/professional/reference/CusDemog.cfm.

Chambers, Aidan. "The Difference of Literature: Writing Now for the Future of Young Readers." *Children's Literature in Education* 24 no. 1 (1993): 1-18.

Cherry, Colin. *On Human Communication: a Review, a Survey, and a Criticism* 2d ed. Cambridge: The MIT Press, 1966.

Chomsky, Noam. *Reflections on Language.* New York: Pantheon, 1975.

Clark, Martin P. *Networks & Telecommunications Design & Operation* 2d ed. Chichester: John Wiley & Sons, 1990.

Clarke, Arthur C. *How the World Was One.* New York: Bantam, 1992.

"Computer Wins Game and Match over Kasparov." Associated Press, 12 May 1997.

Cronin, Blaise, and Kara Overfelt. "E-journals and Tenure." *Journal of the American Society for Information Science* 46 no. 9 (1995): 700-703.

Crooks, Richard M. "Bad Science: the Short Life and Weird Times of Cold Fusion." Book review. *Science* 263 no. 5143 (7 Jan 1994): 105-106.

Crowley, David, and Paul Heyer. *Communication in History.* New York: Longman, 1991.

December, John, and Neil Randall. *The World Wide Web Unleashed.* Indianapolis: Howard Sams, 1994.

de Hartog, Jan. *The Captain.* New York: Atheneum Publishers, 1966.

Denning, Peter J., and Bernard Rous. *The ACM Electronic Publishing Plan,* available on the Internet as http://info.acm.org/pubs/epub_plan.txt, 1994.

Drucker, Peter F. "Infoliteracy." *Forbes ASAP Supplement* 29 August 1994: 104-109.

Duff, Alistair S., David Graig, and David A. McNeil. "A Note of the Origins of the 'Information Society.'" *Journal of Information Science* 22 no. 2 (1996): 117-122.

Duggan, Alfred. *God and My Right*. London: Faber and Faber, 1940. Reproduced with permission of Curtis Brown Ltd., London, on behalf of the estate of Alfred Duggan. Copyright Alfred Duggan.

Earnshaw, Rae, John Vince, and Huw Jones. *Digital Media and Electronic Publishing*. London and San Diego: Academic Press, 1996.

Eisenhart, Douglas M. *Publishing in the Information Age. A New Management Framework for the Digital Era*. Westport CT: Quorum Books, 1994.

Englebart, D. "A Conceptual Framework for Augmentation of Man's Intellect." In *Vistas in Information Handling*. Edited by P. Howerton and D. Weeks. Washington: Spartan Books, 1963, 1-29.

Englebart, D.C., and W.K. English. 'A Research Center for Augmenting Human Intellect. *AFIPS Conference Proceedings, 1968 Fall Joint Computer Conference*, 33, 1968. Montvale,NJ: AFIPS Press, 1968, 395-410.

Feldman, Susan. "DIALOG Update '97." *Information Today*, 14 no. 5 (1997): 10, 12, 30

Flickner, M. and others. "Query by Image and Video Content: the Qbic System." *Computer* 28 no. 9 (Sept 1995): 23-31.

Frankel, Max. "The Daily Digital." *New York Times Magazine*, April 9, 1995, 38.

Fraser, Steven, ed. *The Bell Curve Wars: Race, Intelligence and the Future of America*. New York: Basic Books, 1995.

Gates, Bill, with Nathan Myhrvold, and Peter Rinearson. *The Road Ahead*. New York: Viking Penguin, 1995.

Gore, Al. "Networking the Future: We Need a National 'Superhighway' for Computer Information." *Washington Post*, 15 July 1990, b3.

Greenhouse, Linda. "The Supreme Court: the Opinion; Court, 9-0, Upholds State Laws Prohibiting Assisted Suicide; Protects Speech on Internet." *New York Times,* 27 June 1997, 1.

Grosch, H.R.J. *Computer. Bit Slices from a Life*. Novato,CA: Third Millennium Books, 1991.

Gurrie, Michael, and Patrick J. O'Connor. *Voice-Data Telecommunications Systems: An Introduction to Technology*. Englewood Cliffs: Prentice-Hall, 1985.

Hammerschmidt, D.E. and A.G. Gross. "The Problem of Biomedical

Fraud: A Model for Retrospective and Prospective Action." *Journal of Scholarly Publishing* 27 no. 1 (1995): 3-11.

Harel, Init, Ed. *Constructionist Learning*. Cambridge: MIT Media Laboratory, undated.

Harel, Init, and Seymour Papert, Eds., *Constructionism, Research Reports and Essays, 1985-1990*. Norwood,NJ: Ablex, 1991.

Harpur, Tom. "Information Highway Comes but Wisdom Lingers." *Toronto Star.* January 14, 1996, F4.

Harrison, Teresa M., and Timothy D. Stephen. "The Electronic Journal as the Heart of an Online Scholarly Community." *Library Trends* 43 no. 4 (1995): 592-608.

Hartley, James. "Programmed Instruction 1954-1974: A Review." *Programmed Learning and Educational Technology* 11 no. 6 (1974): 278-291.

Hayakawa, Samuel. *Language in Thought and Action*. New York: Harcourt, Brace & World, 1939.

Hertzberg, Robert. "An All-out Push from Microsoft," *Webweek* 2 no. 1 (January 1996): 1,4.

Hilkevitch, Jon, "CD-ROM Is Taking Frogs' Hearts out of Some Kids' Hands," *Chicago Tribune*. 10 October 1995, 1, 16.

Historical Statistics of the United States, Colonial Times to 1970, Part 2. Washington: U.S. Department of Commerce, 1975.

Hodge, Winston W. *Interactive Television: A Comprehensive Guide to Multimedia Technologies*. New York: McGraw-Hill, 1994.

"IBM Announces 'Net Centric' Computer." *Wall Street Journal* 14 November 1995, B14.

Illich, Ivan. *In the Vineyard of the Text, A Commentary to Hugh's Didascalicon*. Chicago: University of Chicago Press, 1993.

Innis, Harold. *The Bias of Communication*. Toronto: University of Toronto Press, 1951.

The Internet Unleashed. Indianapolis: Howard Sams, 1994.

Jean, Georges. *Writing, The Story of Alphabets and Scripts*. New York: Harry N. Abrams, 1992.

Jenkins, Simon. "The Death of the Written Word." *Journal of Information Science* 21 no. 6 (1995): 407-412. Reprinted with kind permission of Bowker-Saur, a division of Reed Elsevier (UK) Ltd., and the Institute of Information Scientists.

Jensen, Arthur. "Race, Intelligence and Genetics: the Differences Are Real," *Psychology Today* 7 no. 7 (1973): 80-84, 86.

Jensen, Arthur R. "Race." Paper presented at a Symposium of the Institute of Biology on *Racial Variation in Man*. Royal Geographical Society. London, England, 19, 20 September 1974.

Jones, Tim. "Canadian's genetic theories fuel academic freedom war." *Detroit Free Press*, 7 March 1995, 1A.

Joyce, Michael. *Afternoon, a Story*. Cambridge, MA: Eastgate Systems, 1987.

Kennedy, Charles H. *An Introduction to International Telecommunications Law*. Norwood, MA: Artech House, 1996.

Khayyam, Omar. *Rubaiyat of Omar Khayyam*, translated by Edward Fitzgerald. New York: Random House, 1947.

Kiesler, Sara, Jane Siegel, and Timothy McGuire. "Social Psychological Aspects of Computer-mediated Communication." *American Psychologist* 39 no. 10 (1984): 1123-1134

Krol, Ed. *The Whole Internet User's Guide & Catalog*. Sebastopol, CA: O'Reilly & Associates, 1992.

Krumenaker, Larry. "Virtual Libraries, Complete with Journals, Get Real," *Science* 260 no. 5111 (1993): 1066-1067.

Kunhardt, Dorothy. *Pat the Bunny*. Chicago: Golden Books, Western Publishing Co., 1942.

Kurian, George Thomas. *Datapedia of the United States, 1790-2000*. Lanham, MD: Bernan Press, 1994.

LaFollete, M.C. "Stealing into Print: Fraud, Plagiarism, and Misconduct in Scientific Publishing," *Library Quarterly* 64 no. 2 (1994): 221-223.

Lenzner, Robert. "The Reluctant Entrepreneur," *Forbes* (11 Sept. 1995): 162-167, and "Whither Moore's Law?" 167-168.

Lesk, Michael. "Going Digital," *Scientific American* 276 no. 3 (1997): 58-60.

Leslie, J. "Goodbye, Gutenberg. Pixelating Peer Review Is Revolutionizing Scholarly Journals," *Wired* (October 1994): 68-71.

Levarie, Norma. *The Art & History of Books*. New York: Da Capo Press, Undated.

Levy, Stephen. "How the Propeller Heads Stole the Electronic Future. A Future Dominated by Television Is Shattered by the Advent of Internet and Netscape Software," *New York Times Magazine*, 24 September 1995, 58-59.

Lewis, Thomas S.W. *Empire of the Air. The Men Who Made Radio*. Harper Perennial edition. New York: HarperCollins, 1993.

Licklider, J.C.R. "Man-computer Symbiosis." *IRE Transactions on Hu-*

man Factors in Electronics, HFE-1 1 (March 1960): 4-11.

Liddy, E.D., W. Paik, E.S. Yu, and M. McKenna. "Document Retrieval Using Linguistic Knowledge," *RIAO94: Proceedings of the Intelligent Multimedia Information Retrieval Systems and Management Conference*. New York: Rockefeller University, 1995: 106-114.

Logan, Robert K. *The Fifth Language*. Toronto: Stoddart, 1995.

Loring, Brace C., George R. Gamble, and James T. Bond. *Race and Intelligence*. Washington: American Anthropological Association, 1971.

Lyman, Peter. "What Is a Digital Library? Technology, Intellectual Property, and the Public Interest," *Daedalus* 125 no. 4 (1996): 1-33.

Mackie, A."Programmed Learning - a Developing Technique," *Programmed Learning and Educational Technology* 12 no. 4 (1975): 220-228.

Mancall, Jacqueline C., and M. Carl Drott. "Tomorrow's scholars: patterns of facilities use," *School Library Journal* 26 no. 7 (1980): 99-103.

Maran, Stephen P. "Science Is Dandy, but Promotion Can Be Lucrative," *Smithsonian* 22 no. 2 (1991): 72 *et seq.*

Marchionini, Gary. *Information Seeking in Electronic Environments*. Cambridge and New York: Cambridge University Press, 1995.

Marten, A.A. "Development of European Scientific Journal Publishing Before 1850." In *Development of Science Publishing in Europe,* Edited by A.J. Meadows. Amsterdam: Elsevier Science Publishers, 1980, 1-22.

McLuhan, Marshall. *The Gutenberg Galaxy*. Toronto: University of Toronto Press, 1962.

McLuhan, Marshall. *Understanding Media*. New York: McGraw-Hill, 1964.

Meadow, Charles T. *Text Information Retrieval Systems*. San Diego: Academic Press, 1992.

Meadow, Charles T. and Weijing Yuan. "Measuring the Impact of Information: Defining the Concepts," *Information Processing and Management* 33 no. 6 (1997): 697-714.

Meyer, Michael. "A Bad Dream Comes True in Cyberspace." *Newsweek* 127 no. 2 (8 January 1996): 65.

Miller, George A. "Wordnet: an On-line Lexical Database," *International Journal of Lexicography* 3 no. 4 (Winter 1990): 235-312.

Miller, George A. "Wordnet: a Lexical Database for English," *Communications of the ACM* 30 no. 11 (1995): 39-41.

Milliot, Jim. "New Media Helps Spur Sales at McGraw-Hill Cos. Electronic Publishing Boosts Sales, New Multimedia Products for Elementary School Children and Professional Publishing Products," *Publishers Weekly* 243 no. 16 (15 April 1996): 20.

Moghaddam, B. and Alex P. Pentland. "Face Recognition Using View-based and Modular Eigenspaces," *Proceedings of the International Society for Optical Engineering* 2277 (1994): 12-21.

Morton, Frederic. *The Rothschilds, A Family Portrait*. New York: Atheneum, 1962, 48-49.

Mosco, Vincent. *Transforming Telecommunications in Canada*. Ottawa: Canadian Centre for Policy Alternatives, 1990.

Multimedia and CD-ROM Directory, 18th ed, New York: Groves Dictionaries, 1997.

Murray, Charles and Richard J. Herrnstein. *The Bell Curve*. New York: Free Press, 1994.

Negroponte, Nicholas. *Being Digital*. New York: Alfred A. Knopf, 1995.

Nelson, Theodor Holm. "Opening Hypertext: a Memoir." In *Literacy Online*, edited by Myron C. Tuman. Pittsburgh: University of Pittsburgh Press, 1992, 43-57.

Nielsen, Jakob. *Hypertext and Hypermedia*. Boston: Academic Press, 1990.

Norman, Donald A. *The Psychology of Everyday Things*. New York: Basic Books, 1988.

Nyce, James M. and Paul Kahn "A Machine for the Mind: Vannevar Bush's Memex." In: Nyce and Kahn, eds., *From Memex to Hypertext: Vannevar Bush and the Mind's Machine*. Boston: Academic Press, 1991, 39-66.

Oates, Joyce Carol. *Black Water* New York: Dutton, 1992.

Ogden, Frank. *The Last Book You'll Ever Read*. Toronto: Macfarlane Walter & Ross, 1993.

Okerson, A., ed. *Scholarly Publishing on the Electronic Networks. The New Generation: Visions and Opportunities in Electronic Publishing*. Proceedings of the Second Symposium. Washington: Association of Research Libraries, 1993.

Okerson, Ann. "Who Owns Digital Works: Computer Networks Challenge Copyright Law, but Some Proposed Cures May Be as Bad as the Disease," *Scientific American* 275 no. 1 (1996): 80-84

O'Neill, J.E. "The Role of ARPA in the Development of the ARPANET, 1961-1972" *IEEE Annals of the History of Computing* 17 no. 4 (Winter 1995): 76-81.

Paul, Martha. "Common Sense? The Essence of Aroma Therapy: Sniffing Your Way to Vitality," *Chicago Tribune*, 19 November 1995, 4.

Pentland, Alex P. "Smart rooms," *Scientific American* 274 no. 4 (1996): 68-77.

Pinker, Steven. *The Language Instinct*. New York: Harper Perennial, 1994.

"Prestel -- the 'Micro' Database," *Inform* published by the Institute of Information Scientists, 37 (March-April 1981): 3.

Production Screen Life Ltd. *The End of Television*, Witness series, Jean Renaud, producer. Toronto: Canadian Broadcasting Corporation (5 Dec. 1995).

"A Publisher Who Had a Global Electronic Dream," *Financial Times* (London), 16 October 1995, 9.

Quatermain, J.S., and J.C. Hoskins. "Notable Computer Networks," *Communications of the ACM* 29 no. 10 (Oct. 1986): 932-971.

Report of Commissioner to the Court, *The Williams and Wilkins Company v. the United States*, Reprinted by DFS Press, Bethesda, Maryland, 1972.

Rheingold, Howard. *Virtual Reality: The Revolutionary Technology of Computer-Generated Artificial Worlds and How it Promises to Transform Society*. New York: Simon & Schuster, 1992.

Rogers, Everett M. *Diffusion of Innovation,* 3d ed. New York: Free Press, 1983.

Ruby, D. "Terminal behavior," *PC Week* 1 no. 45 (1984): 127-129.

Rushton, J. Philippe, and Anthony F. Bogaert. "Race Differences in Sexual Behavior: Testing an Evolutionary Hypothesis," *Journal of Research in Personality* 21 no. 4 (1987): 529-551.

Rushton, J. Philippe, and R. Travis Osborne. "Genetic and Environmental Contributions to Cranial Capacity in Black and White Adolescents," *Intelligence* 20 no. 1 (1995): 1-13.

Savetz, Kevin M. "Magazines without paper," *Byte* 18 no. 10, (1993): 108.

Seabrook, John. "Gates of the temple," *New Yorker.* (11 December 1995): 78-81.

Sengstack, Jeff. "DVD Drives: Giant CD-ROMs and Movies, Too," *PC World* 14 no. 4 (1996): 50.

Senna, Carl. *The Fallacy of I.Q.* New York: Third Press, 1973.

Shannon, Claude, and Warren Weaver. *The Mathematical Theory of Communication*. Urbana: University of Illinois Press, 1959.

Shapiro, Jeff. *Collaborative Computing, Multimedia Across the Network.*
Boston: AP Professional, 1996.

Shneiderman, Ben. *Designing the User Interface: Strategies for Effective Human-Computer-Interaction.* 2d ed. Reading, MA: Addison-Wesley, 1992.

Snoddy, Raymond. "McGraw-Hill Considers Publishing Move," *Financial Times* (London), 16 October 1995, 19.

Sobel, Robert. *IBM. Colossus in Transition.* New York: A Truman Talley Book, Times Books, 1981.

Soloway, Elliot, and Amanda Pryor. "Next Generation in Human-computer Interaction," *Communications of the ACM* 39 no. 4 (April 1996): 16-18.

"Spielberg's Lament," *New Republic* 198 no. 12 (1988): 7-8.

SRI International, M. Lettor, Network Working Group, Request for Comments 1296. (Since removed from WWW.)

Statistical Abstract of the United States. Lanham, MD: Bernan Press, 1994, 1995, 1996.

Statistical Yearbook 39th Issue. New York: Statistical Division, United Nations, 1994.

Stets, Dan. "Law Experts Hail Internet Ruling as Likely to Last. An Appeal Is Possible. A Reversal Is Doubtful," *Philadelphia Inquirer,* 16 June 1996, E1.

Stoffels, Bob. "Carter Reminisces: How it All Began," *Telephone Engineer & Management* 92 no. 22 (15 Nov. 1988): 48-50.

Sweitzer, James S. "Slice of Life," *The Sciences* 36 no. 2 (1996): 41-43.

Sword, Larry F. "Photocopying and Copyright Law -- Williams & Wilkins Co. v. United States: How Unfair Can `Fair Use' Be?" *Kentucky Law Journal* 63 no. 1 (1975): 256-278.

Tarter, B. "Information Liability: New Interpretations for the Electronic Age," *Computer Law Journal* 11 no. 4 (1992): 481-554.

Taubes, Gary. *Bad Science: The Short Life and Weird Times of Cold Fusion.* New York: Random House, 1991.

Taylor, Susie. "Textbooks from the Terminal (On-demand Custom Printing of College Textbooks)," *Publishers Weekly* 241 no. 24 (13 June 1994): 42-44.

Terkle, Sherry. *The Second Self: Computers and the Human Spirit.* New York: Simon and Schuster, 1984.

"Thomas F. Carter v American Telephone and Telegraph Company, Docket No. 17073, Federal Communications Commission. Re Use of

Carterfone in Message Toll Telephone Service, June 26, 1968," *Public Utilities Reports* 77 PUR 3d, 1968, 417 *et seq.*

Tuman, Myron C., Editor. *Literacy Online.* Pittsburgh: University of Pittsburgh Press, 1992.

UNESCO Division of Statistics on Culture and Communication, *Statistics on Film and Cinema 1955-1977 (Paris: UNESCO, 1981).*

U.S. Economic Review 1996. Encino, CA: Motion Picture Association of America: 1997.

Vonnegut, Kurt. *Player Piano.* New York: Delacorte Press, 1952.

Watson, B.L. "Liability for Failure to Acquire or Use Computers in Medicine." In *Proceedings of the Fifth Annual Symposium on Computer Applications in Medical Care,* (New York: IEEE, 1981): 879-883.

Weidt Group Inc. "CD-ROM Sales Are Growing Rapidly," http://twgi.com/cdsales.htm.

Weizenbaum, Joseph. *Computer Power and Human Reasoning.* San Francisco: W.H. Freeman, 1976.

Wilson, Carl. "Northern Exposure: Canada Fights Cultural Dumping," *The Nation* 262 no. 20 (20 May 1996): 15-18.

Wood, Donald N. and Donald G. Wylie. *Educational Telecommunications.* Belmont, CA: Wadsworth Publishing Co., 1977.

Yanow, Lilli Anne. "Copying. The Reconciliation of Conflicting Interests: an Analysis of Williams & Wilkins Co. V. United States," *Rutgers Journal of Computers and the Law* 3 no. 2 (1974): 328-344.

Young, Jeffrey R. "Classes on the Web," *Chronicle of Higher Education,* 3 November 1995, A27 *et seq.*

Zable, Arnold. *Jewels and Ashes.* New York: Harcourt Brace & Co, 1991.

Zijlstra, Jaco. "The University Licensing Program (Tulip): Electronic Journals in Materials Science," *Microcomputers for Information Management: Global Internetworking for Libraries* 12 nos. 1-2 (1995): 99-112.

Recommended Reading

History

Brooks, John. *Telephone: The First Hundred Years*. New York: Harper & Row, 1976.

Clarke, Arthur C. *How the World Was One*. New York: Bantam, 1992.

Crowley, David, and Paul Heyer. *Communication in History: Technology, Culture, Society*. 2d ed. White Plains, NY: Longman Publishing Group, 1995.

Eisenstein, Elizabeth. *The Printing Press as an Agent of Change*. Cambridge: Cambridge University Press, 1979.

Fisher, David E. and Marshall J. Fisher. *Tube: The Invention of Television*. Orlando: Harcourt Brace, 1997.

Illich, Ivan. *In the Vineyard of the Text, A Commentary to Hugh's Didascalicon*. Chicago: University of Chicago Press, 1993.

Innis, Harold. *Empire and Communication*, 2d ed. Revised by Mary Q. Innis. Toronto: University of Toronto Press, 1972.

Innis, Harold. *The Bias of Communication.* Toronto: University of Toronto Press, 1991.

Lacy, Dan M. *From Grunts to Gigabytes: Communication and Society.* Urbana, IL: University of Illinois Press, 1996.

Lehmann-Haupt, Helmut. *The Book in America: a History of the Making and Selling of Books in the United States.* 2d ed. New York: Bowker, 1951.

Lewis, Thomas S.W. *Empire of the Air.The Men Who Made Radio.* Harper Perennial edition. New York: HarperCollins, 1993.

Logan, Robert K. *The Alphabet Effect.* New York: William Morrow, 1986.

Logan, Robert K. *The Fifth Language.* Toronto: Stoddart, 1995.

Manguel, Alberto. *The History of Reading.* Toronto: Alfred A. Knopf Canada, 1997.

Patterson, Graeme H. *History and Communication: Harold Innis, Marshall McLuhan and the Interpretation of History.* Toronto: University of Toronto Press, 1990.

Rogers, Everett M. *A History of Communication: a Bibliographic Approach.* New York: Free Press, 1994.

Steinberg, S.H. *Five Hundred Years of Printing.* 4th revised ed. New Castle, DE: Oak Knoll Books, 1996.

Mass Media

Dominick, Joseph J. *The Dynamics of Mass Communication.* 5th ed. New York: McGraw-Hill, 1995.

Sloan, W. David. *Perspectives on Mass Communication History.* Hillsdale, NJ: Erlbaum Associates, 1991.

Whitmore, Edward J. *Mediamerica, Mediaworld: Form, Content and Consequences of Mass Communication.* Belmont, CA: Wadsworth, 1995.

Technology and its Applications

Aston, Robert, and Joyce Schwarz. *Multimedia, Gateway to the Next Millenium.* Boston: AP Professional, 1994.

Brand, Stuart. *The Media Lab: Inventing the Future at MIT.* New York: Viking, 1987.

Earnshaw, Rae, John Vince, and Huw Jones. *Digital Media and Elec-*

tronic Publishing. London and San Diego: Academic Press, 1996.

Eisenhart, Douglas M. *Publishing in the Information Age, A New Management Framework for the Digital Era*. Westport, CT: Quorum Books, 1994.

McKnight, C., A. Dillon. and J. Richardson. *Hypertext in Context*. Cambridge: Cambridge University Press, 1991.

Musgrave, James Ray. *The Digital Scribe. A Writer's Guide to Electronic Media*. Boston: AP Professional, 1996.

Nielsen, Jakob. *Hypertext and Hypermedia*. Boston: Academic Press, 1990.

Nunberg, Geoffrey. *The Future of the Book*. Berkeley: University of California Press, 1996.

Peek, Robin, and Gregory B. Newby, eds., *Scholarly Publishing*. Cambridge: The MIT Press, 1996.

Social and Literary Aspects of Information Technology

Birkerts, Sven. *The Gutenberg Elegies: The Fate of Reading in an Electronic Age*. Boston: Faber & Faber, 1994.

de Kerckhove, Derrick. *The Skin of Culture. Investigating the New Electronic Reality*. Toronto: Somerville House, 1995.

McLuhan, Marshall. *The Gutenberg Galaxy. The Making of Typographic Man*. Toronto: University of Toronto Press, 1962.

McLuhan, Marshall. *Understanding Media: the Extensions of Man*. Cambridge: MIT Press, 1994.

McLuhan, Marshall, and Bruce R. Powers. *The Global Village: Transformations in World Life & Media in the 21st Century*. New York: Oxford University Press, 1989.

Mosco, Vincent. *The Pay-per Society: Computers and Communications in the Information Age*. Norwood, NJ: Ablex and Toronto: Garamond, 1989.

Mosco, Vincent, and Janet Wasko, eds. *The Political Economy of Information*. Madison: University of Wisconsin Press, 1988.

Penzias, Arno. *Ideas and Information*. New York: Touchstone - Simon & Schuster, 1989.

Pool, Ithiel de Sola. *The Social Impact of the Telephone*. Cambridge: MIT Press, 1977.

Pool, Ithiel de Sola. *Technologies of Freedom*. Cambridge: Belknap Press of Harvard University Press, 1983.

Rogers, Everett M. *Diffusion of Innovation*. 3d ed. New York: Free Press, 1983.

Stoll, Clifford. *Silicon Snake Oil: Second Thoughts on the Information Highway*. New York: Doubleday, 1995.

Futurology

Negroponte, Nicholas. *Being Digital*. New York: Alfred A. Knopf, 1995.

Ogden, Frank. *The Last Book You'll Ever Read*. Toronto: Macfarlane Walter & Ross, 1993.

Popcorn, Faith. *The Popcorn Report*. New York: HarperCollins, 1992.

Toffler, Alvin. *The Third Wave*. New York: Morrow, 1980.

Toffler, Alvin. *Future Shock*. New York: Random House, 1970.

Index

About the Author

Charles Meadow has more than 40 years of experience in the computer and communications fields. He was educated in mathematics at the University of Rochester and Rutgers University, then became involved very early in database management and information retrieval systems. Later, he focused on the human use of computers, electronic publishing, and communications in general. He is the author of nine books in the computer and communications fields. Two of these were children's books, one of which won Honorable Mention in 1975 in the New York Academy of Science Children's Science Book Awards.

Meadow has worked for the U.S. Government at the Navy Department, National Bureau of Standards, U.S. Office of Science and Technology, and the Atomic Energy Commission. He has industry experience with General Electric Company, IBM, and Dialog Information Services. He joined Drexel University as a Professor in 1974 and the University of Toronto in 1984.